Networking Essentials
Lab Manual

Cisco Networking Academy

D0882381

Cisco Press

221 River St

Hoboken, NJ 07030

Networking Essentials Lab Manual

Cisco Networking Academy

Copyright© 2022 Cisco Systems, Inc.

Published by:
Cisco Press
221 River St
Hoboken, NJ 07030

1 2021

Library of Congress Control Number: 2021916050

ISBN-13: 978-0-13-765926-5
ISBN-10: 0-13-765926-1

Editor in Chief
Mark Taub

Director, Product Line Management
Brett Bartow

Alliance Manager, Cisco Press
Arezou Gol

Executive Editor
James Manly

Managing Editor
Sandra Schroeder

Project Editor
Mandie Frank

Editorial Assistant
Cindy Teeters

Designer
Chuti Prasertsith

Compositor
codeMantra

Proofreader
Charlotte Kughen

CISCO.

Warning and Disclaimer

This book is designed to provide information about networking. Every effort has been made to make this book as complete and as accurate as possible, but no warranty or fitness is implied.

The information is provided on an "as is" basis. The authors, Cisco Press, and Cisco Systems, Inc. shall have neither liability nor responsibility to any person or entity with respect to any loss or damages arising from the information contained in this book or from the use of the discs or programs that may accompany it.

The opinions expressed in this book belong to the author and are not necessarily those of Cisco Systems, Inc.

Trademark Acknowledgments

All terms mentioned in this book that are known to be trademarks or service marks have been appropriately capitalized. Cisco Press or Cisco Systems, Inc., cannot attest to the accuracy of this information. Use of a term in this book should not be regarded as affecting the validity of any trademark or service mark.

This book is part of the Cisco Networking Academy" series from Cisco Press. The products in this series support and complement the Cisco Networking Academy curriculum. For more information on the Cisco Networking Academy or to locate a Networking Academy, please visit netacad.com.

Special Sales

For information about buying this title in bulk quantities, or for special sales opportunities (which may include electronic versions; custom cover designs; and content particular to your business, training goals, marketing focus, or branding interests), please contact our corporate sales department at corpsales@pearsoned.com or (800) 382-3419.

For government sales inquiries, please contact governmentsales@pearson.com

For questions about sales outside the U.S., please contact intlcs@pearson.com

Feedback Information

At Cisco Press, our goal is to create in-depth technical books of the highest quality and value. Each book is crafted with care and precision, undergoing rigorous development that involves the unique expertise of members from the professional technical community.

Readers' feedback is a natural continuation of this process. If you have any comments regarding how we could improve the quality of this book, or otherwise alter it to better suit your needs, you can contact us through email at feedback@ciscopress.com. Please make sure to include the book title and ISBN in your message.

We greatly appreciate your assistance.

CISCO.

Americas Headquarters
Cisco Systems, Inc.
San Jose, CA

Asia Pacific Headquarters
Cisco Systems (USA) Pte. Ltd.
Singapore

Europe Headquarters
Cisco Systems International BV Amsterdam,
The Netherlands

Cisco has more than 200 offices worldwide. Addresses, phone numbers, and fax numbers are listed on the Cisco Website at www.cisco.com/go/offices.

Cisco and the Cisco logo are trademarks or registered trademarks of Cisco and/or its affiliates in the U.S. and other countries. To view a list of Cisco trademarks, go to this URL: www.cisco.com/go/trademarks. Third party trademarks mentioned are the property of their respective owners. The use of the word partner does not imply a partnership relationship between Cisco and any other company. (1110R)

Pearson's Commitment to Diversity, Equity, and Inclusion

Pearson is dedicated to creating bias-free content that reflects the diversity of all learners. We embrace the many dimensions of diversity, including but not limited to race, ethnicity, gender, socioeconomic status, ability, age, sexual orientation, and religious or political beliefs.

Education is a powerful force for equity and change in our world. It has the potential to deliver opportunities that improve lives and enable economic mobility. As we work with authors to create content for every product and service, we acknowledge our responsibility to demonstrate inclusivity and incorporate diverse scholarship so that everyone can achieve their potential through learning. As the world's leading learning company, we have a duty to help drive change and live up to our purpose to help more people create a better life for themselves and to create a better world.

Our ambition is to purposefully contribute to a world where:

- Everyone has an equitable and lifelong opportunity to succeed through learning.

- Our educational products and services are inclusive and represent the rich diversity of learners.

- Our educational content accurately reflects the histories and experiences of the learners we serve.

- Our educational content prompts deeper discussions with learners and motivates them to expand their own learning (and worldview).

While we work hard to present unbiased content, we want to hear from you about any concerns or needs with this Pearson product so that we can investigate and address them.

- Please contact us with concerns about any potential bias at https://www.pearson.com/report-bias.html.

Contents at a Glance

Table of Contents

About This Lab Manual

This is the only authorized Lab Manual for the Cisco Networking Academy Networking Essentials Course.

Networking is at the heart of the digital transformation. The network is essential to many business functions today, including business critical data and operations, cybersecurity, and so much more. A wide variety of career paths rely on the network—so it's important to understand what the network can do, how it operates, and how to protect it.

This is a great course for developers, data scientists, cybersecurity specialists, and other professionals looking to broaden their networking domain knowledge. It's also an excellent launching point for students pursuing a wide range of career pathways—from cybersecurity to software development to business and more. A Networking Academy digital badge is available for the instructor-led version of this course. No prerequisites are required.

You'll learn these core skills:

- Plan and install a home or small business network using wireless technology, then connect it to the Internet.

- Develop critical thinking and problem-solving skills using Cisco Packet Tracer.

- Practice verifying and troubleshooting network and Internet connectivity.

- Recognize and mitigate security threats to a home network. The 37 comprehensive labs in this manual emphasize hands-on learning and practice to reinforce configuration skills.

The Networking Essentials Lab Manual provides you with all the labs and packet tracer activity instructions from the course designed as hands-on practice to develop critical thinking and complex problem-solving skills.

Command Syntax Conventions

The conventions used to present command syntax in this book are the same conventions used in the IOS Command Reference. The Command Reference describes these conventions as follows:

- **Boldface** indicates commands and keywords that are entered literally as shown. In actual configuration examples and output (not general command syntax), boldface indicates commands that are manually input by the user (such as a **show** command).

- *Italic* indicates arguments for which you supply actual values.

- Vertical bars (|) separate alternative, mutually exclusive elements.

- Square brackets ([ ]) indicate an optional element.

- Braces ({ }) indicate a required choice.

- Braces within brackets ([{ }]) indicate a required choice within an optional element.

CHAPTER 1

Communication in a Connected World

 ## 1.5.4 Lab–My Local Network

Objectives

- Record all the different network attached devices in your home or classroom.
- Investigate how each device connects to the network to send and receive information.
- Create a diagram showing the topology of your network.
- Label each device with its function within the network.

Background / Scenario

The path that a message takes from its source to destination can be as simple as a single cable connecting one computer to another or as complex as a network that literally spans the globe. The network infrastructure contains three categories of hardware components:

- End devices
- Intermediary network devices
- Network media

Instructions

Take a close look at the network you have at home or school. Record the network and end-user devices that are connected on the local network.

Sample

Manufacturer	Device	Location	Connection	Media
Apple	iPhone	Mobile	Wireless	WiFi & cell phone
Samsung	Galaxy Smart Phone	Mobile	Wireless	WiFi & cell phone
Cisco	Cable Modem	Home office	Wired	Cable TV coaxial cable and Ethernet cable
Linksys	Wireless Router	Home office	Wired	Ethernet cable
HP	Printer/Scanner	Home office	Wireless	WiFi
Apple	MacBook Air	Mikayla's Room	Wireless	WiFi
Beats by Dre	Headphones	My Room	Wireless	Bluetooth
Microsoft	Xbox	My Room	Wired	Ethernet Cable

Your Local Network

Manufacturer	Device	Location	Connection	Media

Continue the list on a separate page if necessary.

Reflection

1. Are there other electronic devices that are not connected to the local network to share information or resources? What would be the benefit of having these devices online?

2. Which type of connectivity is used most frequently in your local network, wired or wireless?

3. Draw a diagram of your local network. Label each device with a name and location.

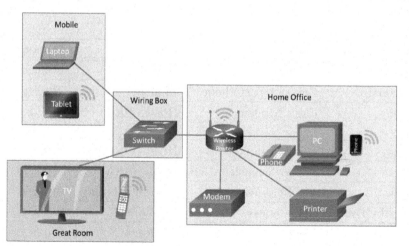

CHAPTER 2

Online Connections

2.2.7 Lab–Determine the IP Address Configuration of a Computer

Objectives

■ In this lab, you will determine the IP address assigned to your computer.

Required Resources

■ 1 PC (Windows 10)

■ Network access

Instructions

Part 1: Determine the IP Address Using the Command Prompt

Step 1: Verify network access.

    a. Open a web browser.

    b. Navigate to any website, such as **www.netacad.com**.

> **Note:** If you do not have any internet access, you can still continue with this lab. Your computer may use Automatic Private Internet Protocol Addressing (APIPA) for IP address information.

Step 2: The command ipconfig.

The **ipconfig** command provides you with the IP address, subnet mask and default gateway.

    a. Click **Start**. Search for **Command Prompt**. Double-click to open a **Command Prompt**.

    b. At the prompt, enter **ipconfig** to determine the IP address assigned to each network adapter on your computer.

```
C:\Users\Student> ipconfig

Ethernet adapter Ethernet0:

    Connection-specific DNS Suffix  . :
    Link-local IPv6 Address . . . . . : fe80::ac29:44a8:6409:c30e%6
    IPv4 Address. . . . . . . . . . . : 192.168.1.11
    Subnet Mask . . . . . . . . . . . : 255.255.255.0
    Default Gateway . . . . . . . . . : 192.168.1.1
```

Questions:

What is the IPv4 address of the computer?

_____

What is the subnet mask of the computer?

_____

What is the default gateway of the computer?

_____

Step 3:   The command **ipconfig /all**.

a.   At the prompt, enter **ipconfig /all** command to view IP configuration on PC-A.

```
C:\Users\Student> ipconfig /all


Ethernet adapter Ethernet0:

    Connection-specific DNS Suffix  . :
    Description . . . . . . . . . . . : Intel(R) 82574L Gigabit Network
                                        Connection
    Physical Address. . . . . . . . . : 00-50-56-B3-E8-C1
    DHCP Enabled. . . . . . . . . . . : Yes
    Autoconfiguration Enabled . . . . : Yes
    Link-local IPv6 Address . . . . . : fe80::ac29:44a8:6409:c30e%6
                                        (Preferred)
    IPv4 Address. . . . . . . . . . . : 192.168.1.11(Preferred)
    Subnet Mask . . . . . . . . . . . : 255.255.255.0
    Lease Obtained. . . . . . . . . . : Sunday, July 24, 2016 9:33:49 AM
    Lease Expires . . . . . . . . . . : Monday, July 25, 2016 10:33:17 AM
    Default Gateway . . . . . . . . . : 192.168.1.1
    DHCP Server . . . . . . . . . . . : 192.168.1.1
    DHCPv6 IAID . . . . . . . . . . . : 50334761
    DHCPv6 Client DUID. . . . . . . . : 00-01-00-01-25-84-55-DE-00-50-
                                        56-B3-E8-C1
    DNS Servers . . . . . . . . . . . : 8.8.8.8
                                        8.8.4.4
    NetBIOS over Tcpip. . . . . . . . : Enabled
```

Questions:

What are the DNS servers for the computer?

_____

What is the MAC address (physical address) of the network adapter?

_____

Is DHCP enabled? If yes, what is the IP address of the DHCP server?

_____

If DHCP is enabled, on what date was the lease obtained? On what date does the lease expire?

Part 2: Test the Network Interface TCP/IP Stack

Step 1: Test TCP/IP stack using the loopback address.

To verify that the TCP/IP protocol is functioning, ping your loopback address (127.0.0.1). Enter the **ping 127.0.0.1** command at the prompt.

```
C:\Users\Student> ping 127.0.0.1
```

Step 2: Test TCP/IP stack using the configured IP address.

You can also ping your IP address. In this example, enter the **ping 192.168.1.11** command at the prompt.

Question:

Record one of the replies from your ping command.

If the ping was not successful, ask your instructor for assistance.

CHAPTER 3

Investigate Networking with Packet Tracer

3.3.3 Packet Tracer–Deploy Devices

Objectives

- In this activity, you will deploy multiple devices.

Required Resources

- Latest version of Packet Tracer

Instructions

A list of device labels is visible in the workspace. We will use various methods to deploy the listed devices.

Note: If you require more help with Packet Tracer, navigate to **Help > Contents** within Packet Tracer.

Part 1: Deploying the Devices

a. Locate the 2911 router in the **Device-Specific Selection Box**.

b. Using your mouse, drag and drop the 2911 router above the **Router0** label in the workspace.

c. Another way to deploy a device is to click the desired device and then click the desired location on the workspace. Click the 1941 router in the **Device-Specific Selection Box** and then click the space above the label **Router1** in the workspace.

d. Use either of the methods to put a 4331 router on the workspace over the **Router2** label.

Part 2: Deploying the Same Type of Devices Multiple Times

If you want to put multiple devices of the same type onto the workspace, the clicking and dragging can become very tedious. To avoid this, you can hold down the <CTRL> key as you click on the devices in the Device-Specific Selection Box.

a. Hold down the <CTRL> key and click the 4321 router in the **Device-Specific Selection Box**.

b. Now click the space above the labels **Router3**, **Router4**, and **Router5**.

c. To cancel the operation, click the **Cancel** icon where the 4321 router was in the **Device-Specific Selection Box**.

Part 3: Copying the Devices

The user can also copy devices on the workspace in two ways.

Method #1: Drag your cursor over the devices that you want to copy.

a. Drag a box over **Router3**. It should appear faded.

b. Hold down the <CTRL> key and drag **Router3** over the label **Copy of Router3**.

c. Repeat this with **Router4** and **Router5**.

Method #2: Hold down the <SHIFT> key and click the devices to be copied.

a. Selecting **Router1** and **Router2** while holding down the <SHIFT> key will again have a faded look.

b. Hold down the <CTRL> key and drag the devices to the space over the label **Copy of Router1** and release.

Packet Tracer
☐ Activity

3.3.4 Packet Tracer–Deploy and Cable Devices

Objectives

- Deploy and Cable Network Devices.

Background / Scenario

You will locate, deploy, and cable multiple types of devices.

Instructions

Part 1: Deploy the Devices

a. Navigate to the **Device-Type Selection** box at the bottom of the screen.

The top row of icons represents categories of devices and the bottom row represents subcategories. Point at the top row of icons slowly and look at the **Label** box between the rows; the names of the categories will appear. Now point at the lower row of icons and you will see their names appear.

b. You will deploy switches and PCs. Point at the lower row of icons until you see one labeled **Switches.** Click the switch icon and you will see the switch devices in the **Device-Specific Selection** box change.

c. Deploy two 2960 switches over the **Switch0** and **Switch1** labels in the workspace. You can drag and drop the switch or select the desired switch and click the desired location in the workspace.

d. Repeat for the end devices. Click the **End Devices** category in the **Device-Type Selection** box and deploy six PCs.

If you are unsure of which device is the PC, just point at the device in the **Device-Specific Selection** box and look at the label area below the devices; it should say PC-PT. (Remember that you do not have to select the PC icon six times to deploy them. There is a shortcut.)

Note: If you need more instructions for deploying the devices, you can refer to a previous Packet Tracer activity or click **Help > Contents >** select **Workspace Basics > Logical Workspace** within Packet Tracer.

Part 2: Cable the Devices

In this part, you will connect the PCs to the switches.

Step 1:   Connect the PCs to the switches.

    a.   Click the category that looks like a lightning bolt labeled **Connections.** In the **Device-Specific Selection** box, there will appear a series of cable types. Select the **Copper Straight-Through** cable type.

    b.   Click the center of the PC0. You will see a pop-up menu appear showing the cable connection types. Select **FastEthernet0** in the pop-up menu.

    c.   With the wire, click **Switch0.** Select **FasEthernet0/1** in the pop-up menu.

    The cable will now be connected, and two blinking link lights will appear, one green and one amber. After a while, the amber light will turn green. You will learn about the color of lights as you progress through the course.

    d.   Repeat the cabling process for the rest of the PC. The connections to be done are listed below:

    PC1 FastEthernet0 to Switch0 FastEthernet0/2

    PC2 FastEthernet0 to Switch0 FastEthernet0/3

    PC3 FastEthernet0 to Switch1 FastEthernet0/1

    PC4 FastEthernet0 to Switch1 FastEthernet0/2

    PC5 FastEthernet0 to Switch1 FastEthernet0/3

Step 2:   Connect the switches.

You will connect the switches together using a copper cross-over cable.

    a.   Select a **Copper Cross-Over** cable. Click **Switch0.** Select **GigabitEthernet0/1** in the pop-up menu.

    b.   With the wire, click **Switch1** and select the same interface from the pop-up menu. The cable will appear and both link lights will be amber but will eventually turn to green after about a minute.

    c.   Save the file as desired.

3.4.3 Packet Tracer–Configure End Devices

Topology

Objectives

- Configure various end devices in Packet Tracer.

Background / Scenario

In this activity, you will construct a simple Packet Tracer network and complete basic configuration of end devices.

Required Resources

Instructions

Part 1: Build the Topology

Step 1:   Create the devices.

Deploy a 2960 switch, two PCs and a server.

If help is required, please refer to previous activities.

Step 2:   Connect the devices.

    a.   Connect FastEthernet0 on PC0 to FastEthernet0/1 on Switch0 with a Copper Straight-Through cable.

    b.   Connect FastEthernet0 on PC1 to FastEthernet0/2 on Switch0 with a Copper Straight-Through cable.

    c.   Connect FastEthernet0 on Server0 to GigabitEthernet0/1 on Switch0 with a Copper Straight-Through cable.

Part 2: Configure Static IP Addresses

Step 1:   Configure the IP address for Server0.

    a.   Click **Server0**.

    b.   Click the **Desktop** tab.

    c.   Click the **IP Configuration** icon.

    d.   Verify the bullet **Static** is selected.

    e.   Enter **192.168.1.1** in the **IP Address** field.

    f.   Enter **255.255.255.0** in the **Subnet Mask** field as needed.

    g.   Close the **IP Configuration** when done.

Step 2:   Configure IP address for the PCs.

    a.   Click **PC0**.

    b.   Click the **Desktop** tab.

    c.   Click the **IP Configuration** icon.

    d.   Verify the bullet **Static** is selected.

    e.   Enter **192.168.1.2** in the **IP Address** field.

    f.   Enter **255.255.255.0** in the **Subnet Mask** field as needed.

    g.   Close the **IP Configuration** when done for PC0.

    h.   Repeat the same procedure for PC1. Use **192.168.1.3** as the IP address for PC1.

Part 3: Verify Connectivity

Step 1:   Verify connectivity via the **Command Prompt**.

    a.   Verify that all the link lights are green.

    b.   Click **PC0**.

    c.   Click the **Desktop** tab.

    d.   Click **Command Prompt** to open the PC command line interface.

    e.   At the prompt, enter **ping 192.168.1.1**.

```
C:\> ping 192.168.1.1
```

If you have done everything correctly, you should see the following output. Your output could vary, but the reply statements should be there. If the replies are not there, try redoing the device configuration to this point.

```
Pinging 192.168.1.1 with 32 bytes of data:

Reply from 192.168.1.1: bytes=32 time<1ms TTL=128
Reply from 192.168.1.1: bytes=32 time<1ms TTL=128
Reply from 192.168.1.1: bytes=32 time<1ms TTL=128
Reply from 192.168.1.1: bytes=32 time=1ms TTL=128

Ping statistics for 192.168.1.1:
Packets: Sent = 4, Received = 4, Lost = 0 (0% loss),
Approximate round trip times in milli-seconds:
Minimum = 0ms, Maximum = 1ms, Average = 0ms
```

    f.   You can also ping **PC1**. Navigate to the **Command Prompt** for PC1 and enter the command **ping 192.168.1.3** at the prompt. The ping should be successful.

    g.   Close the **Command Prompt** when finished.

Step 2:   Verify connectivity via the web browser.

    a.   Click **PC1**.

    b.   Click the **Desktop** tab.

c. Click the **Web Browser** to open the web browser application.

d. Enter **192.168.1.1** in the URL field and click **Go**. The Cisco Packet Tracer webpage should open.

e. Close the web browser when finished.

f. You can also use the web browser application on **PC0** to display the Cisco Packet Tracer webpage. Navigate to PC0. From the **Desktop** tab, open the **Web Browser** and enter **192.168.1.1** in the URL field.

Part 4: Basic Switch Configuration

You will perform some basic configuration on a switch using the **Config** and **CLI** tabs in Packet Tracer.

Step 1: The Config tab.

a. Click **Switch0**.

b. Click the **Config** tab.

Note: The **Config** tab is not always available on physical networking equipment. Some simple devices only have config tabs. The config tab can be useful for basic learning of commands, especially for beginners.

The **Config** tab shows a list of components that can be configured on this device. We are not going to cover what these components are, as that is learned in a networking course, but we will show how to navigate and use the interface.

c. The **Global Settings** enables a user to change the name of a device that displays in the workspace. It also allows for changing the internal name shown at the command line prompt as well as buttons for saving, loading, exporting, and erasing configuration files.

Double-click in the **Hostname** dialog box to highlight the word Switch. Enter **Central** to replace Switch as the hostname. Packet Tracer will display the IOS commands necessary to accomplish the name change in the **Equivalent IOS Commands** box. The commands displayed should be as follows:

```
Switch> enable
Switch# configure terminal
Enter configuration commands, one per line. End with CNTL/Z.
Switch(config)# hostname Central
Central(config)#
```

These would be the commands used from the command line interface, or CLI, to change the hostname. If you did not know how to do this from the CLI, the **Config** tab would show the necessary commands.

d. Click the **FastEthernet0/1** under the Interface heading to configure the FastEthernet0/1 interface.

In the **Equivalent IOS Commands**, the command **interface FastEthernet0/1** is displayed in the **Equivalent IOS Commands** box.

Step 2:   The CLI tab.

    a.   Select the **CLI** tab to switch to the CLI interface. Notice that the same commands that were in the **Equivalent IOS Commands** box are listed in the CLI window.

    b.   At the prompt, enter **shutdown**.

```
Central(config-if)# shutdown
Central(config-if)#
%LINK-5-CHANGED: Interface FastEthernet0/1, changed state to
administratively down
%LINEPROTO-5-UPDOWN: Line protocol on Interface FastEthernet0/1, changed
state to down
Central(config-if)#
```

This command just shuts down the interface down from the command line.

    c.   Navigate to the Workspace. Notice that the link lights for the connection between PC0 and Switch0 are red. Because the interface on the switch was shut down, the connection is no longer active and shows red.

    d.   Save and close the activity, then exit Packet Tracer if desired.

3.5.1 Packet Tracer–Create a Simple Network

Addressing Table

Device	Interface	IP Address	Subnet Mask	Default Gateway
PC	Ethernet0	DHCP		192.168.0.1
Wireless Router	LAN	192.168.0.1	255.255.255.0	N/A
	Internet	DHCP		
cisco.com Server	Ethernet0	209.165.200.225	255.255.255.224	N/A
Laptop	Wireless0	DHCP		192.168.0.1

Objectives

In this activity, you will build a simple network in Packet Tracer.

- Part 1: Build a Simple Network
- Part 2: Configure the End Devices and Verify Connectivity

Instructions

Part 1: Build a Simple Network

In this part, you will build a simple network by deploying and connecting the network devices.

Step 1:   Add network devices to the workspace.

In this step, you will add a PC, laptop, and a cable modem to the Logical Workspace.

Using the device selection box, add the devices to the workspace, add the following devices to the workspace. The category and sub-category associated with the device are listed below:

- **PC:** End Devices > End Devices > PC
- **Laptop:** End Devices > End Devices > Laptop
- **Cable Modem:** Network Devices > WAN Emulation > Cable Modem

Step 2:   Change display names of the nework devices.

  a. To change the display names of the network devices, click the device icon on the Logical workspace.

  b. Click the **Config** tab in the **Device Configuration Window**.

  c. Enter the new name of the newly added device into the **Display Name** field according to the Addressing Table.

Step 3:   Add the physical cabling between devices on the workspace.

Using the device selection box, add the physical cabling between devices on the workspace.

a. The PC will need a copper straight-through cable to connect to the wireless router. Select the copper straight-through cable in the device selection box and attach it to the **FastEthernet0** interface of the PC and the **GigabitEthernet 1** interface of the wireless router.

b. The wireless router will need a copper cross-over cable to connect to the cable modem. Select the copper cross-over cable in the device-selection box and attach it to the Internet interface of the wireless router and the **Port 1** interface of the cable modem.

c. The cable modem will need a Coaxial cable to connect to the Internet cloud. Select the **Coaxial cable** in the device-selection box and attach it to the **Port 0** interface of the cable modem and the **Coaxial 7** interface of the Internet cloud.

Part 2: Configure the End Devices and Verify Connectivity

Step 1:   Configure the PC.

You will configure the PC for the wired network in this step.

a. Click the **PC**. In the **Desktop** tab, navigate to **IP Configuration** to verify that DHCP is enabled and the PC has received an IP address. Close the **IP Configuration** window when done.

b. In the **Desktop** tab, click **Command Prompt**.

c. Verify that the PC has received an IPv4 address by issuing the **ipconfig /all** command from the prompt. The PC should receive an IPv4 address in the 192.168.0.x range.

d. Test connectivity to the cisco.pka server from the PC. From the command prompt, issue the command **ping cisco.pka**. It may take a few seconds for the ping to return. Four replies should be received.

Step 2:   Configure the laptop.

In this step, you will configure the laptop to access the wireless network.

a. Click **Laptop**, and select the **Physical** tab.

b. In the **Physical** tab, you will need to remove the Ethernet copper module and replace it with the Wireless WPC300N module.

   1) Power off **Laptop** by clicking the power button on the side of the laptop.

   2) Remove the currently installed Ethernet copper module by clicking on the module on the side of the laptop and dragging it to the **MODULES** pane on the left of the laptop window.

   3) Install the Wireless WPC300N module by clicking it in the **MODULES** pane and dragging it to the empty module port on the side of the laptop.

   4) Power on the **Laptop** by clicking the Laptop power button again.

c. With the wireless module installed, connect the laptop to the wireless network. Click the **Desktop** tab and select **PC Wireless.**

d. Select the **Connect** tab. The wireless network **HomeNetwork** should be visible in the list of wireless networks. Select the **HomeNetwork.** Click **Connect.**

e. Close **PC Wireless.** Select **Web Browser** in the **Desktop** tab.

f. In the web browser, navigate to **cisco.pka.**

CHAPTER 4

Build a Simple Network

 ### 4.5.7 Lab–Build a Simple Network

Topology

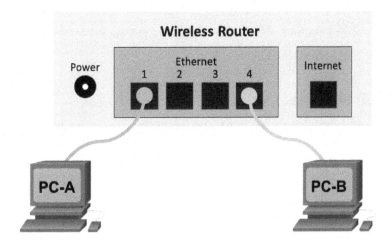

Addressing Table

Device	Interface	IP Address	Subnet Mask
PC-A	NIC	192.168.1.10	255.255.255.0
PC-B	NIC	192.168.1.11	255.255.255.0

Objectives

Part 1: Set Up the Network Topology (Ethernet only).

- Identify cables and ports for use in the network.
- Cable a physical lab topology.

Part 2: Configure PC Hosts.

- Enter static IP address information on the LAN interface of the hosts.
- Verify that PCs can communicate using the **ping** utility.

Background / Scenario

Networks are constructed of three major components: hosts, switches, and routers. In this lab, you will build a simple network with two hosts and a wireless router with at least two switchports. You will apply IP addressing for this lab to the PCs to enable communication between these two devices. Use the ping utility to verify connectivity.

Required Resources

- 1 wireless router with at least two switchports
- 2 PCs running Windows with wired network cards installed
- 2 Ethernet patch cables

Instructions

Part 1: Set Up the Network Topology (Ethernet only)

In Part 1, you will cable the devices together according to the network topology.

Step 1: Power on the devices.

Power on all devices in the topology.

Step 2: Connect the PCs to the switch.

  a. Connect one end of an Ethernet cable to the NIC port on PC-A. Connect the other end of the cable to a switchport on the wireless router. After connecting the PC to the switchport, you should see the light for the switchport turn amber and then green, indicating that PC-A has been connected correctly.

  b. Repeat the same procedure for PC-B.

Step 3: Visually inspect network connections.

After cabling the network devices, take a moment to carefully verify the connections to minimize the time required to troubleshoot network connectivity issues later.

Part 2: Configure PC Hosts

In this lab, all the network configurations are done on a Windows 10 PC.

Step 1: Configure static IP address information on the PCs.

  a. To configure the Network Settings on PC-A, click **Start**, then click **Settings**.

  b. In the **Settings** window click **Network & Internet**.

  c. In the left pane select **Ethernet**, then click **Change adapter options**.

  d. The **Network Connections** window displays the available network interfaces on the PC. Right-click the **Ethernet0** interface and select **Properties**.

    e. Select the **Internet Protocol Version 4 (TCP/IPv4)** option and then click **Properties.**

> **Note:** You can also double-click **Internet Protocol Version 4 (TCP/IPv4)** to display the **Properties** window.

    f. Click the **Use the following IP address** radio button to manually enter an IP address, subnet mask, and default gateway. Type in the IP address 192.168.1.10 and the subnet mask 255.255.25.0.

> **Note:** In the above example, the IP address and subnet mask have been entered for PC-A. The default gateway has not been entered because the router is not configured. Refer to the Addressing Table on page 1 for PC-B's IP address information.

    g. After all the IP information has been entered, click **OK**. Click **OK** on the **Ethernet0 Properties** window to assign the IP address to the LAN adapter. Click **Close** to close the **Ethernet0 Status** window.

    h. Repeat the previous steps to enter the IP address information on PC-B.

Step 2: Verify PC settings and connectivity.

Use the **Command Prompt** to verify the PC settings and connectivity.

    a. From PC-A, click **Start** and search for **Command Prompt**.

    b. The **Command Prompt** window is where you can enter commands directly to the PC and view the results of those commands. Verify your PC settings by using the **ipconfig /all** command. This command displays the PC hostname and the IP address information.

    c. Type **ping 192.168.1.11**.

    Question:

    Were the ping results successful?

> If the ping was unsuccessful, there is good chance that **Windows Firewall** is blocking ICMP echo requests (ping). Click **Start > Settings > Network & Internet > Ethernet > Windows Firewall** to turn it off. For security purpose, you should return the firewall back to the original state when you are finished with the lab.

> **Note:** If you did not get a reply from PC-B, try to ping PC-B again. If you still do not get a reply from PC-B, try to ping PC-A from PC-B. If you are unable to get a reply from the remote PC, ask your instructor to help you troubleshoot the problem.

4.5.8 Lab–Trace a Route

Objectives

- Determine network connectivity to a destination host.
- Trace a route to a remote server using tracert.

Background / Scenario

Data travels from a source end device to a destination device. Route tracing software lists the path traversed by this data.

This route tracing software is typically executed at the command line as:

```
tracert <destination network name or end device address>
```

(Microsoft Windows systems)

or

```
traceroute <destination network name or end device address>
```

(UNIX, Linux systems, and Cisco devices, such as switches and routers)

Both **tracert** and **traceroute** determine the route taken by packets across an IP network.

The **tracert** (or **traceroute**) tool is often used for network troubleshooting. By showing a list of routers traversed, the user can identify the path taken to reach a particular destination on the network or across internetworks. Each router represents a point where one network connects to another network and through which the data packet was forwarded. The number of routers traversed is known as the number of hops the data traveled from source to destination.

Command-line based route tracing tools are usually embedded with the operating system of the end device. This activity should be performed on a computer that has Internet access and access to a command line.

Required Resources

- PC with Internet access

Instructions

Part 1: Determine Network Connectivity to a Destination Host

To trace the route to a distant network, the PC must have a working connection to the Internet. Use the ping command to test whether a host is reachable. Packets of information are sent to the remote host with instructions to reply. The PC measures whether each packet receives a response, and how long it takes for those packets to cross the network.

a. Navigate to the Command Prompt, enter **ping www.cisco.com** to determine if it is reachable.

b. Now ping one of the Regional Internet Registry (RIR) websites located in different parts of the world to determine if it is reachable:

Africa: **www.afrinic.net**

Australia: **www.apnic.net**

South America: **www.lacnic.net**

North America: **www.arin.net**

Note: At the time of writing, the European RIR www.ripe.net does not reply to ICMP echo requests.

The website you selected will be used in the next part for use with the **tracert** command.

Part 2: Trace a Route to a Remote Server Using Tracert

After you use **ping** to determine if your chosen websites are reachable, you will use **tracert** to determine the path to reach the remote server. Look closely at each network segment that is crossed.

Each hop in the **tracert** results displays the routes that the packets take when traveling to the final destination. The PC sends three ICMP echo request packets to the remote host. Each router in the path decrements the time to live (TTL) value by 1 before passing it onto the next system. To decrement is to count down. When the decremented TTL value reaches 0, the router sends an ICMP Time Exceeded message back to the source with its IP address and the current time. When the final destination is reached, an ICMP echo reply is sent to the source host.

Step 1:   Trace a route to www.cisco.com.

At the prompt, trace the route to **www.cisco.com**.

```
C:\Users\User1> tracert www.cisco.com

Tracing route to e144.dscb.akamaiedge.net [23.67.208.170]
over a maximum of 30 hops:

  1      1 ms     <1 ms     <1 ms   192.168.1.1
  2     14 ms      7 ms      7 ms   10.39.0.1
  3     10 ms      8 ms      7 ms   172.21.0.118
  4     11 ms     11 ms     11 ms   70.169.73.196
  5     10 ms      9 ms     11 ms   70.169.75.157
  6     60 ms     49 ms        *    68.1.2.109
  7     43 ms     39 ms     38 ms   Equinix-DFW2.netarch.akamai.com
                                    [206.223.118.102]
  8     33 ms     35 ms     33 ms   a23-67-208-170.deploy.akamaitechnologies.com
                                    [23.67.208.170]

Trace complete.
```

In this example, the source host sends three ICMP echo request packets to the first hop (192.168.1.1) with the TTL value of 1. When the router 192.168.1.1 receives the echo request packets, it decrements the TTL value to 0. The router sends an ICMP Time Exceeded message back to the source. This process continues until the source host sends the last three ICMP echo request packets with TTL values of 8 (hop number 8 in the output above), which is the final destination. After the ICMP echo request packets arrive at the final destination, the router responds to the source with ICMP echo replies.

For hops 2 and 3, these IP addresses are private addresses. These routers are the typical setup for point-of-presence (POP) of the ISP. The POP devices connect users to an ISP network.

Step 2: Trace a route to a RIR web site.

a. Now perform a **tracert** to one of the RIR web sites from the previous part.

Africa:	**www.afrinic.net**
Australia:	**www.apnic.net**
South America:	**www.lacnic.net**
North America:	**www.arin.net**

b. A web-based **whois** tool is found at http://whois.domaintools.com/. It can be used to determine the domains traveled from the source to destination.

Question:

List the domains below from your **tracert** results using a web-based whois tool, such as http://whois.domaintools.com/.

CHAPTER 5

Communication Principles

 ## 5.1.4 Lab–My Protocol Rules

Objectives

- Relate computer network protocols to the rules that you use every day for various forms of communication.
- Define the rules that govern how you send and interpret text messages.
- Explain what would happen if the sender and receiver did not agree on the details of the protocol.

Background / Scenario

Before beginning to communicate with each other, we establish rules or agreements to govern the conversation. These rules, or protocols, must be followed for the message to be successfully delivered and understood. Among the protocol characteristics that govern successful human communication are:

- An identified sender and receiver
- Agreed upon method of communicating
- Common language and grammar
- Speed and timing of delivery
- Confirmation or acknowledgment requirements

The techniques that are used in network communications share these fundamentals with human conversations.

Instructions

Think about the commonly accepted protocol standards for sending text messages to your friends. Fill out the chart on the next page with some of the rules that you follow when texting with friends and others. The first row has been filled in as an example.

Your Text Messaging Protocol

Protocol Requirement	What does this mean?	How is it implemented in your protocol?
An identified sender and receiver	How do you know who the text message is from? How does the person on the other end know the message is delivered to you? Is it going to an individual or a group?	In text messaging, the sender and receiver are usually identified by telephone number, username, or nickname. Group texts can be sent to a predefined group or new groups created on demand.
Agreed upon method of communicating	Do we send text only? Do we send pictures back & forth? What about using smileys and emoji?	
Common language and grammar	Do we use acronyms? Is slang acceptable? What is the native language of the participants?	
Speed and timing of delivery	What determines how soon the recipient gets the message? How quickly to we expect to receive a response?	
Confirmation or acknowledgment requirements	How do you know that the message was received? How do you know that the conversation is finished?	

Reflection

1. Now that you have documented the protocols that you use when sending and reading text messages, do you think that these protocols would be the same if you were texting with friends or with your parents and teachers? Explain your answer.

2. What do you think that the consequences would be if there were no agreed-upon protocol standards for different methods of communications?

3. Share your protocol rules with your classmates. Are there differences between your protocols and theirs? If so, could these differences result in misunderstanding of the messages?

5.4.5 Lab–Determine the MAC Address of a Host

Addressing Table

Device	Interface	IP Address	Subnet Mask
PC	VLAN 1	192.168.1.2	255.255.255.0

Objectives

- Determine the MAC address of a Windows computer on an Ethernet network using the **ipconfig /all** command.

- Analyze a MAC address to determine the manufacturer.

Background / Scenario

Every computer on an Ethernet local network has a Media Access Control (MAC) address that is burned into the Network Interface Card (NIC). Computer MAC addresses are usually displayed as 6 sets of two hexadecimal numbers separated by dashes or colons (example: 15-EF-A3-45-9B-57). The **ipconfig /all** command displays the computer MAC address. You may work individually or in teams.

Required Resources

- PC running Windows 10 with at least one Ethernet network interface card (NIC)

- Connectivity to the Internet

Instructions

Part 1: Locating the MAC Address on a Computer

In this part of the lab, you will determine the MAC address of a computer using the Windows **ipconfig** command.

Step 1:   Display information for the command **ipconfig / all**.

   a.   Right-click on the **Start** button and select **Command Prompt**.

   b.   Enter the **ipconfig /all** command at the command prompt.

Step 2: Locate the MAC (physical) address(es) in the output from the **ipconfig /all** command.

Use the table below to fill in the description of the Ethernet adapter and the physical (MAC) Address:

Description	Physical Address

Question:

How many MAC addresses did you discover in your PC?

Part 2: Analyzing the Parts of a MAC Address

Every Ethernet network interface has a physical address assigned to it when it is manufactured. These addresses are 48 bits (6 bytes) long and are written in hexadecimal notation. MAC addresses are made up of two parts. One part of the MAC address, the first 3 bytes, represents the vendor who manufactured the network interface. This part of the MAC is called the OUI (Organizationally Unique Identifier). Each vendor who wants to make and sell Ethernet network interfaces must register with the IEEE in order to be assigned an OUI.

The second part of the address, the remaining 3 bytes, are the unique ID for the interface. All MAC addresses that begin with the same OUI must have unique values in the last 3 bytes.

In this example, the physical MAC address for the Ethernet LAN interface is D4-BE-D9-13-63-00.

Manufacturer OUI	Unique Identifier for the Interface	Vendor Name
D4-BE-D9	13-63-00	Dell Incorporated

Step 1: List MAC addresses discovered by you and your classmates in the previous part.

List the 3-byte Manufacturer OUI and the 3-byte unique interface identifier. You will fill in the Vendor name in the table below.

Manufacturer OUI	Unique Identifier for the Interface	Vendor Name
D4-BE-D9	13-63-00	Dell Incorporated

Manufacturer OUI	Unique Identifier for the Interface	Vendor Name

Step 2:   Look up the vendors who are the registered owners of the OUIs that you listed in the table.

   a.   Wireshark.org provides an easy-to-use lookup tool at https://www.wireshark.org/tools/oui-lookup.html. Use this tool or use the Internet to search for other ways to identify an OUI.

   b.   Use the information that you found to update the vendor column in the chart in Step 1.

   Question:

   How many different vendors did you discover?

Reflection

1.   Why might a computer have more than one MAC address?

2.   The sample output from the **ipconfig /all** command shown previously had only one MAC address. Suppose the output was from a computer that also had wireless Ethernet capability. How might the output change?

3.   Try connecting and disconnecting the network cable(s) to your network adapter(s) and use the **ipconfig /all** command again. What changes do you see? Does the MAC address still display? Will the MAC address ever change?

4.   What are other names for the MAC address?

CHAPTER 6

Network Design and the Access Layer

 ## 6.2.4 Lab–View Wireless and Wired NIC Information

Objectives

- Part 1: Identify and Work with PC NICs.
- Part 2: Identify and Use the System Tray Network Icons.

Background / Scenario

This lab requires you to determine the availability and status of the network interface cards (NICs) on the PC. Windows provides a number of ways to view and work with your NICs.

In this lab, you will access the NIC information of the PC and change the status of these cards.

Required Resources

- 1 PC (Windows 10 with two NICs, wired and wireless, and a wireless connection)
- A wireless router

Instructions

Part 1: Identify and Work with PC NICs

In Part 1, you will identify the NIC types in the PC. You will explore different ways to extract information about these NICs and how to activate and deactivate them.

Note: This lab was performed using a PC running the Windows 10 operating system. You should be able to perform the lab with another Windows operating systems version. However, menu selections and screens may vary.

Step 1: Use network connections.

You will verify which network connections are available.

a. Right-click **Start** and select **Network Connections**.

b. The **Network Connections** window displays the list of NICs available on this PC. Look for your Local Area Connection and Wireless Network Connection adapters in this window.

Note: If the Network status page is displayed, click **Change adapter options** to navigate to the Network Connections window.

Note: Other types of network adapters, such as Bluetooth Network connection and Virtual Private Network (VPN) adapter, may also be displayed in this window.

Step 2:   Work with your wireless NIC.

Verify the wireless network connection settings.

a.   Right-click a **Wireless Network Connection.** The first option displays if your wireless NIC is enabled or disabled. If your wireless NIC is disabled, you will have an option to **Enable** it.

b.   Verify that the wireless network is connected. If not, click **Connect/Disconnect** to connect to the desired network. Click **Status** to open the **Wireless Network Connection Status** window.

Questions:

What is the Service Set Identifier (SSID) for the wireless router of your connection?

_____

What is the speed of your wireless connection?

_____

c.   Click **Details** to display the Network Connection Details window.

Question:

What is the MAC address of your wireless NIC?

_____

d.   Open a Command Prompt and enter **ipconfig /all.**

```
C:\Users\Bob> ipconfig /all
```

Notice the information displayed is similar to the **Network Connection Details** window information. When you have reviewed the details, click **Close** to return to the **Wireless Network Connection Status** window.

e.   Return to the **Wireless Network Connection Status** window. Click **Wireless Properties** to open the **Wireless Network Properties** window of the Home-Net network.

f.   You should always use wireless security whenever available. To verify (or configure) the wireless security options click on the **Security** tab.

The window displays the type of security and encryption method enabled. You can also enter (or change) the security key in this window. Close all windows.

Step 3:   Work with your wired NIC.

We will now verify the wired network connection settings.

a.   Open the **Network Connections** window by right-clicking **Windows Start > Network Connections.**

Note: If the Network status page is displayed, click **Change adapter options** to navigate to the **Network Connections** window.

    b. Select and right-click the **Local Area Connection** option to display the drop-down list. If the NIC is disabled, enable it.

    c. Click the **Status** option to open the **Local Area Connection Status** window. This window displays information about your wired connection to the LAN.

    d. Click **Details…** to view the address information for your LAN connection.

    e. Open a Command Prompt and enter **ipconfig /all**. Find your Local Area Connection information and compare this with the information displayed in the **Network Connection Details** window.

    f. Close all windows on your desktop.

Part 2: Identify and Use the System Tray Network Icons

In Part 2, you will use the network icons in your system tray to display the available networks.

    a. The bottom right-hand corner of the Windows 10 screen contains the system tray. Move your mouse to display the system tray.

    b. If you hover over the network icon in the system tray, it displays the currently connected networks.

    c. Click the wireless network icon, and it displays the wired and wireless network SSIDs that are in range of your wireless NIC.

    d. Right-click the wireless network icon, and it displays a troubleshooting option and to open the **Network and Sharing Center** window.

    e. Click the **Open Network and Sharing Center** option.

    f. The Network and Sharing Center is a central window that displays information about the active network or networks, the network type, and the type of access.

Reflection

Why would you activate more than one NIC on a PC?

6.4.8 Lab–View Captured Traffic in Wireshark

Topology

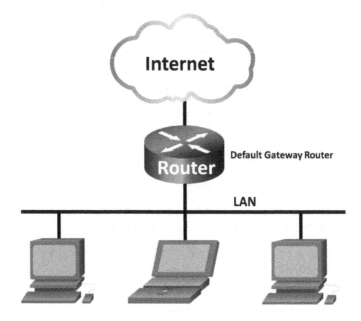

Objectives

Part 1: Download and Install Wireshark.

Part 2: Capture and Analyze ARP Data in Wireshark.

- Start and stop data capture of ping traffic to remote hosts.
- Locate the IPv4 and MAC address information in captured PDUs.
- Analyze the content of the ARP messages exchanged between devices on the LAN.

Part 3: View the ARP cache entries on the PC.

- Access the **Windows Command Prompt.**
- Use the Windows **arp** command to view the local ARP table cache on the PC.

Background / Scenario

Address Resolution Protocol (ARP) is used by TCP/IP to map a Layer 3 IPv4 address to a Layer 2 MAC address. When an Ethernet frame is transmitted on the network, it must have a destination MAC address. To dynamically discover the MAC address of a known destination, the source device broadcasts an ARP request on the local network. The device that is configured with the destination IPv4 address responds to the request with an ARP reply and the MAC address is recorded in the ARP cache.

Every device on the LAN maintains its own ARP cache. The ARP cache is a small area in RAM that holds the ARP responses. Viewing an ARP cache on a PC displays the IPv4 address and the MAC address of each device on the LAN with which the PC has exchanged ARP messages.

Wireshark is a software protocol analyzer, or "packet sniffer" application, used for network troubleshooting, analysis, software and protocol development, and education. As data streams travel back and forth over the network, the sniffer "captures" each protocol data unit (PDU) and can decode and analyze its content according to the appropriate protocol specifications.

Wireshark is a useful tool for anyone working with networks and can be used with most labs in the Cisco courses for data analysis and troubleshooting. This lab provides instructions for downloading and installing Wireshark, although it may already be installed. In this lab, you will use Wireshark to capture ARP exchanges on the local network.

Required Resources

- 1 PC (Windows 10)

- Internet access

- Additional PC(s) on a local-area network (LAN) will be used to reply to **ping** requests. If no additional PCs are on the LAN, the default gateway address will be used to reply to the **ping** requests.

Instructions

Part 1: Download and Install Wireshark

Wireshark has become the industry standard packet-sniffer program used by network engineers. This open source software is available for many different operating systems, including Windows, Mac, and Linux.

If Wireshark is already installed on your PC, you can skip Part 1 and go directly to Part 2. If Wireshark is not installed on your PC, check with your instructor about your academy's software download policy.

Step1:   Download Wireshark.

    a.   Wireshark can be downloaded from www.wireshark.org.

    b.   Click **Download**.

    c.   Choose the software version you need based on your PC's architecture and operating system. For instance, if you have a 64-bit PC running Windows, choose **Windows Installer (64-bit)**.

    d.   After making the selection, the download should start. Click **Save File** if prompted. The location of the downloaded file depends on the browser and operating system that you use. For Windows users, the default location is the **Downloads** folder.

Step 2:   Install Wireshark.

    a.   The downloaded file is named **Wireshark-win64-x.x.x.exe**, where **x** represents the version number. Double-click the file to start the installation process.

b. Respond to any security messages that may display on your screen. If you already have a copy of Wireshark on your PC, you maybe prompted to uninstall the old version before installing the new version. It is recommended that you remove the old version of Wireshark prior to installing another version. Click **Yes** to uninstall the previous version of Wireshark.

c. If this is the first time installing Wireshark, or after you have completed the uninstall process, you will navigate to the Wireshark Setup wizard. Click **Next**.

d. Continue advancing through the installation process. Click **I Agree** when the **License Agreement** window displays.

e. Keep the default settings on the **Choose Components** window and click **Next**.

f. Choose your desired shortcut options and click **Next**.

g. You can change the installation location of Wireshark, but unless you have limited disk space, it is recommended that you keep the default location.

h. To capture live network data, Npcap must be installed on your PC. If your installed version of Npcap is older than the version that comes with Wireshark, it is recommended that you allow the newer version to be installed by clicking the **Install Npcap x.x.x** (version number) check box. Click **Next**.

i. A separate window opens up for Npcap Setup. Click **I Agree** to in the **Npcap License Agreement** window. In the **Installation Options** window, leave all the checkboxes unselected and click **Install** to install Npcap. Click **Next** when finished. Click **Finish** to close the wizard.

j. The installation of USBPcap is not necessary for this course. It is only required if you are planning to capture USB traffic. Click **Install** to start the installation.

Note: Because USBPcap is experimental, make sure that you have created a system restore point before the installation of USBPcap.

---

k. Wireshark starts installing its files and a separate window displays with the status of the installation. Click **Next** when the installation is complete.

l. Click **Finish** to complete the Wireshark install process. If the installation process is stalled, verify that the Npcap installation is finished. Click **Next** to continue.

m. Reboot the PC to finish the installation.

Part 2: Capture and Analyze Local ARP Data in Wireshark

In Part 2 of this lab, you will ping another PC on the LAN and capture ARP requests and replies in Wireshark. You will also look inside the frames captured for specific information. This analysis should help to clarify how packet headers are used to transport data to their destination.

Step 1:   Retrieve your PC's interface addresses.

For this lab, you will need to retrieve your PC's IPv4 address and the MAC address.

a.   Navigate to a Command Prompt window. Type **ipconfig /all** at the prompt.

b. Note which network adapter that the PC is using to access the network. Record your PC interface's IPv4 address and MAC address (physical address).

```
C:\Users\Student> ipconfig /all

<output omitted>

Wireless LAN adapter Wireless Network Connection:

   Connection-specific DNS Suffix  . :
   Description . . . . . . . . . . . : Intel(R) Centrino(R) Advanced-N
                                       6205
   Physical Address. . . . . . . . . : A4-AE-31-AD-78-4C
   DHCP Enabled. . . . . . . . . . . : Yes
   Autoconfiguration Enabled . . . . : Yes
   Link-local IPv6 Address . . . . . : fe80::f9e7:e41d:a772:f993%11
                                       (Preferred)
   IPv4 Address. . . . . . . . . . . : 192.168.1.8(Preferred)
   Subnet Mask . . . . . . . . . . . : 255.255.255.0
   Lease Obtained. . . . . . . . . . : Thursday, August 04, 2016
                                       05:35:35 PM
   Lease Expires . . . . . . . . . . : Friday, August 05, 2016
                                       05:35:35 PM
   Default Gateway . . . . . . . . . : 192.168.1.1
   DHCP Server . . . . . . . . . . . : 192.168.1.1
   DHCPv6 IAID . . . . . . . . . . . : 245648945
   DHCPv6 Client DUID. . . . . . . . : 00-01-00-01-1B-87-BF-52-A4-4E-
                                       31-AD-78-4C
   DNS Servers . . . . . . . . . . . : 192.168.1.1
   NetBIOS over Tcpip. . . . . . . . : Disabled
```

c. Ask a team member for their PC's IPv4 address and give your PC's IPv4 address to them. Do not provide them with your MAC address at this time.

Question:

Record the IPv4 addresses of the default gateway and the other PCs on the LAN.

Step 2: Start Wireshark and begin capturing data.

a. On your PC, click **Start** and type **Wireshark**. Click **Wireshark Desktop App** when it appears in the search results window.

Note: Alternatively, your installation of Wireshark may also provide a Wireshark Legacy option. This displays Wireshark in the older but widely recognized GUI. The remainder of this lab was completed using the newer Desktop App GUI.

---

b. After Wireshark starts, select the network interface that you identified with the **ipconfig** command. Enter **arp** in the filter box. This selection configures Wireshark to only display packets that are part of the ARP exchanges between the devices on the local network. After you have selected the correct interface and entered the filter information, click **Start capturing packets** (shark fin icon) to begin the data capture.

Information will start scrolling down the top section in Wireshark. Each line represents a message being sent between a source and destination device on the network.

c. In a Command Prompt window, ping the default gateway to test the connectivity to the default gateway address that was identified Part 2, Step 1.

```
C:\Users\Student> ping 192.168.1.1

Pinging 192.168.1.1 with 32 bytes of data:
Reply from 192.168.1.1: bytes=32 time=7ms TTL=64
Reply from 192.168.1.1: bytes=32 time=2ms TTL=64
Reply from 192.168.1.1: bytes=32 time=1ms TTL=64
Reply from 192.168.1.1: bytes=32 time=6ms TTL=64

Ping statistics for 192.168.1.1:
    Packets: Sent = 4, Received = 4, Lost = 0 (0% loss),
Approximate round trip times in milli-seconds:
    Minimum = 1ms, Maximum = 7ms, Average = 4ms
```

d. Ping the IPv4 addresses of other PCs on the LAN that were provided to you by your team members.

Note: If your team member's PC does not reply to your pings, this may be because their PC firewall is blocking these requests. Ask your instructor for assistance to disable the PC firewall if necessary.

---

e. Stop capturing data by clicking **Stop Capture** (red square icon) on the toolbar.

Step 3:   Examine the captured data.

In Step 3, examine the data that was generated by the **ping** requests of your team member's PC. Wireshark data is displayed in three sections:

1) The top section displays the list of PDU frames captured with a summary of the IPv4 packet information listed.

2) The middle section lists PDU information for the frame selected in the top part of the screen and separates a captured PDU frame by its protocol layers.

3) The bottom section displays the raw data of each layer. The raw data is displayed in both hexadecimal and decimal form.

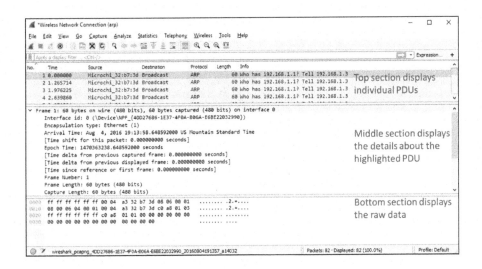

a. Click one of the ARP frames in the top section that has your PC MAC address as the source address in the frame and "broadcast" as the destination of the frame.

b. With this PDU frame still selected in the top section, navigate to the middle section. Click the arrow to the left of the Ethernet II row to view the Destination and Source MAC addresses.

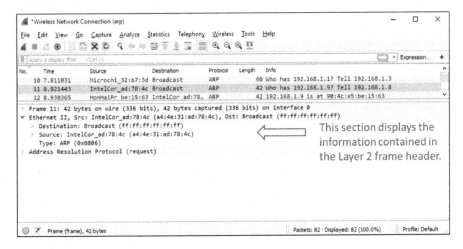

Question:

Does the Source MAC address match your PC's interface?

c. Click the arrow to the left of the Address Resolution Protocol (request) row to view the content of the ARP request.

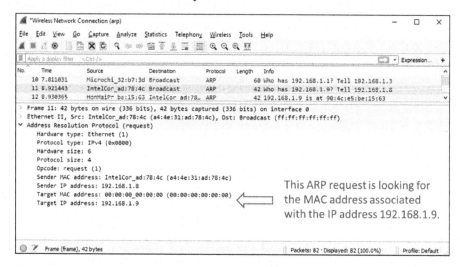

Step 4:   Locate the ARP response frame that corresponds to the ARP request that you highlighted.

a. Using the Target IPv4 address in the ARP request, locate the ARP response frame in the upper section of the Wireshark capture screen.

Question:

What is the IPv4 address of the Target device in your ARP request?

_____

b. Highlight the response frame in the upper section of the Wireshark output. You may have to scroll the window to find the response frame that matches the Target IPv4 address identified in the previous step. Expand the Ethernet II and Address Resolution Protocol (response) rows in the middle section of the screen.

Questions:

Is the ARP response frame a broadcast frame?

_____

What is the destination MAC address of the frame?

_____

Is this the MAC address of your PC?

_____

What MAC address is the source of the frame?

_____

c. Verify with your team member that the MAC address matches the MAC address of their PC.

Part 3: Examine the ARP cache entries on the PC

After the ARP reply is received by the PC, the MAC Address to IPv4 address association is stored in cache memory on the PC. These entries will stay in memory for a short period of time (from 15 to 45 seconds), then, if they are not used within that time, they will be removed from cache.

a. Open a command prompt window on the PC. At the prompt, enter **arp –a** and press enter.

```
C:\Users\Student> arp -a


Interface: 192.168.1.8 --- 0xb
    Internet Address        Physical Address       Type
    192.168.1.1             00-37-73-ea-b1-7a      dynamic
    192.168.1.9             90-4c-e5-be-15-63      dynamic
    192.168.1.13            a4-4e-31-ad-78-4c      dynamic
    224.0.0.5               01-00-5e-00-00-05      static
    224.0.0.6               01-00-5e-00-00-06      static
    224.0.0.22              01-00-5e-00-00-16      static
    224.0.0.252             01-00-5e-00-00-fc      static
    224.0.0.253             01-00-5e-00-00-fd      static
    239.255.255.250         01-00-5e-7f-ff-fa      static
    255.255.255.255         ff-ff-ff-ff-ff-ff      static
```

The output of the **arp –a** command displays the entries that are in the cache on the PC. In the example, the PC has entries for the default gateway (192.168.1.1) and for two PCs that are located on the same LAN (192.168.1.9 and 192.168.1.13).

Question:

What is the result of executing the **arp –a** command on your PC?

b. The **arp** command on the Windows PC has another functionality. Enter **arp /?** at the command prompt and press enter. The **arp** command options enable you to view, add and remove ARP table entries if necessary.

Questions:

Which option deletes an entry from the ARP cache?

What would be the result of issuing the **arp –d** * command?

Reflection

1. What is a benefit of keeping ARP cache entries in memory on the source computer?

   _____

2. If the destination IPv4 address is not located on the same network as the source host, what MAC address will be used as the destination target MAC address in the frame?

   _____

CHAPTER 7

Routing Between Networks

7.1.4 Lab–IPv4 Addresses and Network Communication

Topology

Objectives

- Build a simple peer-to-peer network and verify physical connectivity.
- Assign various IPv4 addresses to hosts and observe the effects on network communication.

Background / Scenario

In this lab, you will build a simple peer-to-peer network using two PCs and an Ethernet crossover cable. You will assign various IPv4 addresses to the hosts and determine the effects on their ability to communicate.

Required Resources

- 2 PCs (Windows 10 with at least one wired Ethernet NIC on each PC)
- 1 Ethernet crossover cable to connect the PCs (provided by instructor)

Instructions

Part 1: Create a Peer-to-Peer Network

Step 1: Connect the PCs to create a peer-to-peer network.

    a. Obtain an Ethernet crossover cable provided by the instructor to connect the two PCs.

        Note: (optional lab setup) The PCs may be connected to a switch using two straight-through cables. The following instructions assume you are using a crossover cable.

---

    b. Plug one end of the cable into the Ethernet NIC of PC-A. Plug the other end of the cable into the Ethernet NIC of PC-B. As you insert the cable, you should hear a click which indicates that the cable connector is properly inserted into the port.

Step 2:   Verify physical connectivity.

a.   After the Ethernet crossover cable is connected to both PCs, take a close look at each Ethernet port. A link light (usually green or amber) indicates that physical connectivity has been established between the two NICs. Try unplugging the cable from one PC then reconnecting it to verify that the light goes off then back on.

b.   On PC-A, right-click **Start** and select **Network Connections.**

c.   If there was a problem connecting the network cable, Ethernet0 will read **Network cable unplugged** and will have a red X in the icon. If so, troubleshoot by repeating Steps 1 and 2. You may also want to ask your instructor to confirm that you are using an Ethernet crossover cable.

Step 3:   Configure IPv4 Settings on PC-A and PC-B.

a.   Configure manual IPv4 addressing on PC-A and PC-B so that they can communicate using TCP/IP. On PC-A, right click the **Ethernet0 icon** and click **Properties.**

b.   In the Ethernet0 Properties window, select **Internet Protocol Version 4 (TCP/IPv4)** and click the **Properties** button.

c.   Select the **Use the following IPv4 address** and enter the IPv4 address **192.168.1.1** and subnet mask **255.255.255.0**. Click **OK > Close** to exit the Ethernet0 Properties window.

d.   Repeat steps 3a–3c for PC-B using the IPv4 address **192.168.1.2** and subnet mask **255.255.255.0.**

Step 4:   Verify IPv4 connectivity between the two PCs.

Note: To test TCP/IP connectivity between the PCs, Windows Firewall must be disabled temporarily on both PCs. Windows Firewall should be re-enabled after the tests have been completed. To access Windows Firewall, click **Settings > Network & Internet > Ethernet > Windows Firewall > Turn Windows Firewall on or off.**

---

a.   Now that the two PCs are physically connected and configured correctly with IPv4 addresses, we need to make sure they can communicate with each other. The **ping** command is a simple way to accomplish this task.

b.   Navigate to the Command Prompt on PC-A. In a Command Prompt on PC-A, ping **192.168.1.2**. A successful **ping** will verify network connectivity and that PC-A can communicate with PC-B.

```
C:\Users\Student> ping 192.168.1.2

Pinging 192.168.1.2 with 32 bytes of data:
Reply from 192.168.1.2: bytes=32 time=4ms TTL=64
Reply from 192.168.1.2: bytes=32 time=1ms TTL=64
Reply from 192.168.1.2: bytes=32 time=1ms TTL=64
Reply from 192.168.1.2: bytes=32 time<1ms TTL=64

Ping statistics for 192.168.1.2:
    Packets: Sent = 4, Received = 4, Lost = 0 (0% loss),
Approximate round trip times in milli-seconds:
    Minimum = 0ms, Maximum = 4ms, Average = 1ms
```

c. Repeat this procedure and **ping 192.168.1.1** from PC-B.

d. Close the command prompt on both PCs.

Part 2: Modify IPv4 Addresses

Step 1: Change the IPv4 address for PC-B.

a. On PC-B, right-click on **Start** and select **Network Connections**, and right-click the **Ethernet0** icon. Choose **Properties** from the pull-down menu.

b. Select **Internet Protocol Version 4 (TCP/IPv4)**. Click **Properties**.

c. Change the logical IPv4 address for PC-B from 192.168.1.2 to **192.168.2.2** and leave the subnet mask set to 255.255.255.0.

d. Click **OK**, which will close the **Internet Protocol Version 4 (TCP/IPv4)** window. Click the **Close** button to exit the **Ethernet0 Properties** window.

Step 2: Test network connectivity between the 2 PCs.

a. From PC-B, right-click **Start**, and select **Command Prompt**.

b. At the command prompt, type **ping 192.168.2.2** and press **Enter**.

Questions:

Was it successful? Explain.

_____

What type of networking device would allow the PCs to communicate even though they are on different networks?

_____

Step 3: Change IPv4 address for PC-A.

a. On PC-A, right-click **Start** and select **Network Connections**, and right-click the **Ethernet0** icon. Choose **Properties** from the pull-down menu.

b. Select **Internet Protocol Version 4 (TCP/IP)**. Click **Properties**.

c. Change the logical IPv4 address for PC-A from 192.168.1.1 to **192.168.2.99** and leave the subnet mask set to **255.255.255.0**. Click **OK** to close the **Internet Protocol Version 4 (TCP/IP)** window. Click **Close** to exit the **Ethernet0 Properties** window.

The two PCs are still on the same physical Ethernet network.

Question:

Are they on the same logical IPv4 network now?

_____

Step 4:   Test network connectivity between the 2 PCs.

    a.   On PC-B, repeat step 4b to access the Windows command prompt.

    b.   At the command prompt, **ping 192.168.2.99.**

    Question:

    Was it successful? Explain.

Step 5:   (Optional – Use only if the Firewall was originally ENABLED).

To ensure that the PC is protected from unauthorized access, re-enable the Windows Firewall.

To access Windows Firewall, click **Settings > Network & Internet > Ethernet > Windows Firewall > Turn Windows Firewall on or off.**

7.3.3 Packet Tracer–Observe Data Flow in a LAN

Objectives

- Develop an understanding of the basic functions of Packet Tracer.
- Create/model a simple Ethernet network using 3 hosts and a switch.
- Observe traffic behavior on the network.
- Observe data flow of ARP broadcasts and pings.

Instructions

Part 1: Create a Logical Network Diagram with 3 PCs and a Switch

During an activity, to ensure that the instructions always remain visible, click the "top" check box in the instruction box window.

The bottom left hand corner of the Packet Tracer screen displays the icons that represent device categories or groups, such as Routers, Switches, or End Devices.

Moving the cursor over the device categories will show the name of the category in the box. To select a device, first select the device category. When the device category is selected, the options within that category appear in the box next to the category listings. Select the device option that is required.

a. Select **End Devices** from the options in the bottom left-hand corner. Drag and drop 3 generic PCs onto your design area.

b. Select **Switch** from the options in the bottom left-hand corner. Add a 2960 switch to your prototype network by dragging it onto your design area.

c. Select **Connections** from the bottom left-hand corner. Choose a copper straight-through cable type. Click the first host (PC0) and assign the cable to the **FastEthernet0** connector. Click the switch (Switch0) and select the connection **FastEthernet0/1** for PC0.

d. Repeat Step c for PC1 and PC2. Select **FastEthernet0/2** on the Switch0 for PC1 and **FastEthernet0/3** for PC2.

There should be green dots at both ends of each cable connection after the network has converged. If not, double check the cable type selected.

Part 2: Configure Host Names and IP Addresses on the PCs

a. Click **PC0**. Select the **Config** tab. Change the PC Display Name to **PC-A**. Select the **FastEthernet** tab on the left and add **192.168.1.1** as the IP address and **255.255.255.0** as the subnet mask. Close PC-A when done.

b. Click **PC1**. Select the **Config** tab. Change the PC Display Name to **PC-B**. Select the **FastEthernet** tab on the left and add **192.168.1.2** as the IP address and **255.255.255.0** as the subnet mask. Close PC-B when done.

c. Click **PC2**. Select the **Config** tab. Change the PC Display Name to **PC-C**. Select the **FastEthernet** tab on the left and add **192.168.1.3** as the IP address and **255.255.255.0** as the subnet mask. Close PC-C when done.

Part 3: Observe the Flow of Data from PC-A to PC-C by Creating Network Traffic

a. Switch to **Simulation Mode** in the bottom right-hand corner.

b. Click **Edit Filter** in the Edit List Filter area. In the event list filter, click **All/None** to deselect every filter. Click **Edit Filter**. Select **ARP** and **ICMP** filters under the IPv4 tab.

c. Select a **Simple PDU** by clicking the closed envelope in the upper toolbar. With the envelop icon, click **PC-A** to establish the source. Click **PC-C** to establish the destination.

Note: Notice that two envelopes are now positioned beside PC-A. One envelope is ICMP, while the other is ARP. The Event List in the Simulation Panel will identify exactly which envelope represents ICMP and which represents ARP.

---

d. Select **Play** from the Play Controls in the Simulation Panel. You can speed up the simulation using the **Play Speed Slider**. The **Play Speed Slider** is located below **Play** inside the Simulation Panel. Dragging the button to the right will speed up the simulation, while dragging it to the left will slow down the simulation.

e. Observe the path of the ICMP and ARP envelopes. Click **View Previous Event** to continue when the buffer is full.

f. Click **Reset Simulation** in the Simulation Panel. Notice that the ARP envelope is no longer present. This has reset the simulation but has not cleared any configuration changes or dynamic table entries, such as ARP table entries. The ARP request is not necessary to complete the ping because PC-A already has the MAC address in the ARP table.

g. Click **Capture then Forward** inside the Simulation Panel. The ICMP envelope will move from the source to the switch and stop. The **Capture then Forward** option allows you to move the simulation one step at a time. Continue selecting **Capture then Forward** until you complete the event.

h. Click the **Power Cycle Device** on the bottom left, above the device icons.

i. An error message will appear asking you to confirm reset. Click **Yes**. Now both the ICMP and ARP envelopes are present again. The power cycle will clear any configuration changes not saved and will clear all dynamic table entries, such as the ARP and MAC table entries.

j. Exit Simulation Mode by clicking **Realtime** and allow the network to converge.

k. After the network has converged, enter **Simulation Mode**.

Part 4: View ARP Tables on Each PC

a. Click **Play** to repopulate the ARP table on the PCs. Click **View Previous Event** when the buffer is full.

b. Click **Inspect** (magnifying glass) in the upper tool bar.

c. With the magnifying glass, click **PC-A**. Select **ARP Table** in the pop-up menu. Notice that PC-A has an ARP entry for PC-C. View the ARP tables for PC-B and PC-C as well. Close all ARP table windows.

d. Click **Select** in the upper tool bar.

e. Click PC-A and select the **Desktop** tab.

f. Select the **Command Prompt** and enter the command **arp -a** to view the ARP table from the desktop view. Close the PC-A configuration window.

```
C:\> arp -a

Internet Address        Physical Address        Type

192.168.1.3             0003.e406.e430          dynamic
```

g. Examine the ARP tables for PC-B and PC-C. Close the **Command Prompt** window when finished.

7.3.4 Lab–Connect to a Wireless Router

Topology

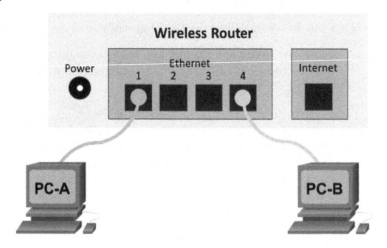

Addressing Table

Device	Interface	IP Address	Subnet Mask	Default Gateway
PC-A	NIC	192.168.10.2	255.255.255.0	192.168.10.1
PC-B	NIC	192.168.10.3	255.255.255.0	192.168.10.1

Objectives

- Connect a PC to a wireless router using Ethernet cable.
- Configure the PC with an appropriate IPv4 address.
- Verify the PC configuration using a Command Prompt.

Background / Scenario

For a PC to communicate in the local network and the Internet, it must be connected to a network device.

Required Resources

- 2 PCs (Windows 10 with wired Ethernet NIC on each PC)
- 1 wireless router
- 2 straight-through Ethernet cables

Instructions

Part 1: Connecting to the Network

Step 1: Connect the end devices.

    a. Power on the end devices and wireless router as needed.

    b. Connect two PCs to the wireless router using Ethernet cables.

Step 2: Configure the PCs with addressing information.

    a. In PC-A, right-click the **Start** button then select **Network Connections**. If the Network status window opens, click **Change adapter options** to navigate to **Network Connections**.

    b. Select the wired network connection. Right-click the desired wired network connection and select **Properties**.

    c. Open the **Internet Protocol Version 4 (TCP/IPv4)** option to open the TCP/IP properties window.

    d. You will enter an IPv4 address configuration consisting of an IPv4 address, a subnet mask, and default gateway address. To enter the address information, click the **Use the following IP address** button.

    e. In the IPv4 address field, enter **192.168.10.2**. In the subnet mask field, enter **255.255.255.0**. In the default gateway field, enter **192.168.10.1**. The DNS server information is not necessary at this time.

    f. When finished, click **OK** to return to the **Internet Protocol (TCP/IPv4) Properties** window. Click **OK** to apply the changes.

    After the changes are applied, you will be returned to the **Network Connections** window.

    g. Because the two computers are on the same network, their IPv4 addresses will be similar, and their subnet masks and default gateways will be identical. Perform the same procedures on PC-B to assign an IPv4 address, subnet mask, and default gateway using the following information:

    Question:

    Why do you think the IPv4 addresses are different, but the subnet masks and default gateways are the same?

Step 3:   Verify connectivity between the two PCs.

    a.   In a **Command Prompt** window, enter **ipconfig /all** to verify the configured IPv4 address and the default gateway from the previous step for both PCs.

    b.   From the command prompt on PC-A, test connectivity with PC-B by entering **ping 192.168.10.3**.

    c.   The pings should be successful. If the pings are not successful, perform the appropriate troubleshooting steps, such as checking the cabling and checking your IPv4 address, subnet mask, and default gateway assignments.

Part 2: Some Useful Windows Networking Commands

Within the Command Prompt window, you can access a few useful networking utilities besides the ones you have seen already: **ipconfig**, **ping**, and **tracert**.

Step 1:   The **hostname** command

Besides the **ipconfig /all** command, you can print the name of the current host using the **hostname** command.

At a command prompt, enter **hostname**.

```
C:\Users\Student> hostname
DESKTOP-3FR7RKA
```

Step 2:   The **getmac** command

Sometimes you just need to quickly list all the MAC addresses for the NIC on the PC. The **getmac** command could be quite useful.

At a command prompt, enter **getmac**.

```
C:\Users\Student> getmac

Physical Address      Transport Name
==================    =============================================================
00-50-56-B3-E8-C1     \Device\Tcpip_{B0D0B9B3-8A23-4B59-B930-323792047552}
02-00-4C-4F-4F-50     \Device\Tcpip_{31C9748F-BCD8-4C76-8AEB-D4BC482341B8}
```

Step 3:   The **route** command

The **route** command allows you to manipulate the network routing table. For this example, you will print out the routes.

```
C:\Users\Student> route print
===========================================================================
Interface List
  6...00 50 56 b3 e8 c1 ......Intel(R) 82574L Gigabit Network Connection
  3...02 00 4c 4f 4f 50 ......Microsoft KM-TEST Loopback Adapter
  1........................Software Loopback Interface 1
  7...00 00 00 00 00 00 00 e0 Microsoft ISATAP Adapter
  5...00 00 00 00 00 00 00 e0 Microsoft ISATAP Adapter #2
===========================================================================
```

```
IPv4 Route Table
===========================================================================
Active Routes:
Network Destination        Netmask          Gateway       Interface  Metric
          0.0.0.0          0.0.0.0      192.168.1.1  192.168.1.11      10
        127.0.0.0        255.0.0.0        On-link      127.0.0.1     306
        127.0.0.1  255.255.255.255        On-link      127.0.0.1     306
  127.255.255.255  255.255.255.255        On-link      127.0.0.1     306
      169.254.0.0      255.255.0.0        On-link   169.254.5.92     266
     169.254.5.92  255.255.255.255        On-link   169.254.5.92     266
  169.254.255.255  255.255.255.255        On-link   169.254.5.92     266
      192.168.1.0    255.255.255.0        On-link   192.168.1.11     266
     192.168.1.11  255.255.255.255        On-link   192.168.1.11     266
    192.168.1.255  255.255.255.255        On-link   192.168.1.11     266
        224.0.0.0        240.0.0.0        On-link      127.0.0.1     306
        224.0.0.0        240.0.0.0        On-link   192.168.1.11     266
        224.0.0.0        240.0.0.0        On-link   169.254.5.92     266
  255.255.255.255  255.255.255.255        On-link      127.0.0.1     306
  255.255.255.255  255.255.255.255        On-link   192.168.1.11     266
  255.255.255.255  255.255.255.255        On-link   169.254.5.92     266
===========================================================================
Persistent Routes:
  None


IPv6 Route Table
===========================================================================
Active Routes:
 If Metric Network Destination      Gateway
  1    306 ::1/128                   On-link
  6    266 fe80::/64                 On-link
  3    266 fe80::/64                 On-link
  3    266 fe80::8050:173e:a32b:55c/128
                                     On-link
  6    266 fe80::ac29:44a8:6409:c30e/128
                                     On-link
  1    306 ff00::/8                  On-link
  6    266 ff00::/8                  On-link
  3    266 ff00::/8                  On-link
===========================================================================
Persistent Routes:
  None
```

CHAPTER 8

The Internet Protocol

8.1.2 Packet Tracer–Connect to a Web Server

Objectives

- Observe how packets are sent across the Internet using IP addresses.

Instructions

Part 1: Verify Connectivity to the Web Server

a. Open the source host command prompt window. Select PC0.

b. Select the **Desktop Tab > Command Prompt.**

c. Verify connectivity to the web server. At the command prompt, ping the IP address of the web server by entering **ping 172.33.100.50.**

```
PC> ping 172.33.100.50

Pinging 172.33.100.50 with 32 bytes of data:

Reply from 172.33.100.50: bytes=32 time=0ms TTL=127
Reply from 172.33.100.50: bytes=32 time=0ms TTL=127
Reply from 172.33.100.50: bytes=32 time=0ms TTL=127
Reply from 172.33.100.50: bytes=32 time=0ms TTL=127

Ping statistics for 172.33.100.50:
Packets: Sent = 4, Received = 3, Lost = 1 (25% loss),
Approximate round trip times in milli-seconds:
Minimum = 0ms, Maximum = 0ms, Average = 0ms
```

A reply verifies connectivity from the client to the destination web server. The reply may time out initially while devices load and ARP is performed.

d. Close the command prompt window only, by selecting the **x** within the command prompt window. Be sure to leave the PC0 configuration window open.

Part 2: Connect to the Web Server via the Web Client

a.   In the **Desktop** tab on PC0, select **Web Browser**.

b.   Enter **172.33.100.50** into the URL and click **Go**. The web client will connect to the web server via the IP address, and open the web page.

Question:

What messages did you see after the web page has finished loading?

8.2.7 Lab–Use a Calculator for Binary Conversions

Objectives

- Switch among Windows Calculator modes.
- Use Windows Calculator to convert between decimal and binary.
- Use Windows Calculator to determine the number of hosts in a network with powers of 2.

Background / Scenario

When working with networking devices, a network technician should understand binary and decimal numbers. In this lab, you will use the Windows Calculator application to convert between these numbering systems. You will also use the "powers" function to determine the number of hosts that can be addressed based on the number of bits available.

Required Resources

- PC (Windows 10)

Instructions

Part 1: Using Windows Calculator

Step 1: Access Windows Calculator and determine mode of operation.

    a. Open the Windows Calculator application. Click **Start**, type **Calculator**. Select **Calculator** in the results.

    Question:

    What mode is the Calculator in?

    b. Click the **Open Navigation** icon (≡) located in the upper left-hand corner. The Windows calculator supports different mode of calculations.

    Question:

    List the five modes for the calculator.

Step 2: Switch between modes.

    a. To switch between calculator modes, click the **Open Navigation** icon (≡) and select desired mode.

    b. Practice switching between calculator modes to see which options they provide.

    Question:

    Briefly explain the function of each mode.

Part 2: Binary System

Step 1:   Convert between number systems.

a.   Select the **Programmer** mode calculator.

b.   The **Programmer** calculator supports four numbering systems — HEX
     (Hexadecimal), DEC (Decimal), OCT (Octal), and BIN (Binary).

Questions:

Which number system is currently active?

_____

Which numbers on the number pad are active in Decimal mode?

_____

c.   Click **BIN.**

Questions:

Which numbers on the number pad are now active?

_____

Why do you think the other numbers are grayed out?

_____

d.   Click **DEC.** Using your mouse, click on the number **1** followed by the number **5**
     on the number pad. The decimal number 15 has now been entered.

e.   Now click **BIN.**

Question:

What happened to the number 15 listed in the textbox at the top of the
window?

_____

f.   Enter the number 220 and select BIN.

Question:

What is the binary equivalent of 220?

_____

g.   Clear the binary value representing 220 in the window. From Binary mode, type
     in the following binary number: **11001100.** Select the **DEC.**

Question:

What is the decimal equivalent to the binary number of 11001100?

_____

Step 2:   Practice the conversion between binary and decimal numbering system.

Questions:

Convert the following decimal numbers to binary.

Decimal	Binary
86	
175	
204	
19	

Convert the following binary numbers to decimal.

Binary	Decimal
1100 0011	
0010 1010	
0011 1000	
1001 0011	

Part 3: Network Addresses

Step 1:   Convert host IP addresses.

a.   Computer hosts usually have two addresses, an Internet Protocol (IP) address and an Ethernet Media Access Control (MAC) address. For the benefit of humans, the IP address is normally represented in dotted decimal notation, such as 192.168.10.2. Each of the decimal octets in the address or a mask can be converted to 8 binary bits. Remember that the computer only understands binary bits.

Question:

If all 4 octets were converted to binary, how many bits would there be?

_____

b.   IP addresses are normally shown with four decimal numbers ranging from 0 to 255 and separated by a period.

Question:

Convert the 4 parts of the IP address 192.168.10.2 to binary.

Decimal	Binary
192	
168	
10	
2	

Step 2:   Convert host IP subnet masks.

Subnet masks, such as 255.255.255.0, are also represented as dotted decimal. A subnet mask will always consist of four 8-bit octets, each one represented as a decimal number. With the exception of decimal 0 (all 8 binary zeros) and decimal 255 (all 8 binary ones), each octet will have some number of ones on the left and some number of zeros on the right.

Questions:

Convert the 8 possible decimal subnet octet values to binary.

Decimal	Binary
0	
128	
192	
224	
240	
248	
252	
254	
255	

Convert the four parts of the subnet mask 255.255.255.0 to binary.

Decimal	Binary
255	
255	
255	
0	

Step 3:   Manipulate powers of 2 to determine the number of hosts on a network.

a. Binary numbers use two digits, 0 and 1. When you calculate how many hosts can be on a subnetwork, you use powers of two because binary is being used. As an example, we have a subnet mask that leaves six bits in the host portion of the IP address. In this case, the number of hosts on that network is 2 to the 6th power minus 2 (because you need a number to represent the network and a number that can be used to reach all the hosts—the broadcast address). The number 2 is always used because we are working in binary. The number 6 is the number of bits that are used for the host bits.

b. Change the calculator view to **Scientific** mode. Input the number **2**. Select the **xy** key on the calculator, the key which raises a number to a power. Input the number **6**. To complete the operation, click on the = key, press **Enter** on the keyboard, or press the = key on the keyboard. The number 64 appears in the output. To subtract two, click on the minus (-) key and then the **2** key followed by the = key. The number 62 appears in the output. This means 62 hosts could be utilized.

Questions:

Using the previously described process, determine the number of hosts if the following number of bits are used for host bits.

No. of Bits Used for Hosts	No. of Hosts
5	
14	
24	
10	

Using a similar technique as learned previously, determine what 10 to the 4th power equals.

c. Close the Windows Calculator application.

Reflection

List one other thing for which you might use the Windows Calculator scientific mode. It does not have to be related to networking.

CHAPTER 9

Dynamic Addressing with DHCP

9.2.5 Packet Tracer–Configure DHCP on a Wireless Router

Objectives

- Connect 3 PCs to a wireless router.
- Change the DHCP setting to a specific network range.
- Configure clients to obtain their address via DHCP.

Background / Scenario

A home user wants to use a wireless router to connect 3 PCs. All 3 PCs should obtain their address automatically from the wireless router.

Instructions

Part 1: Set Up the Network Topology

a. Add three generic PCs.

b. Connect each PC to an Ethernet port on the wireless router using straight-through cables.

Part 2: Observe the Default DHCP Settings

a. After the amber lights have turned green, click **PC0**. Click the **Desktop** tab. Select **IP Configuration**. Select **DHCP** to receive an IP address from the **DHCP Enabled Router**.

Question:

Record the IP address of the default gateway:

_____

b. Close the **IP Configuration** window.

c. Open a web browser.

d. Enter the IP address of the default gateway recorded earlier into the URL field. When prompted, enter the username **admin** and password **admin**.

e. Scroll through the **Basic Setup** page to view default settings, including the default IP address of the wireless router.

f. Notice that DHCP is enabled, and note the starting address of the DHCP range and the range of addresses available to clients.

Part 3: Change the Default IP Address of the Wireless Router

    a. Within the **Router IP Settings** section, change the IP address to: **192.168.5.1**.

    b. Scroll to the bottom of the page and click **Save Settings**.

    c. If it is done correctly, the web page will display an error message. Close the web browser.

    d. Click **IP Configuration** to renew the assigned IP address. Click **Static**. Click **DHCP** to receive new IP address information from the wireless router.

    e. Open the web browser and enter the IP address **192.168.5.1** in the URL field. When prompted, enter the username **admin** and password **admin**.

Part 4: Change the Default DHCP Range of Addresses

    a. Notice the **DHCP Server Start IP Address** is updated to the same network as the Router IP.

    b. Change the **Starting IP Address** from 192.168.5.100 to **192.168.5.126**.

    c. Change the **Maximum Number of Users** to **75**.

    d. Scroll to the bottom of the page and click **Save Settings**. Close the web browser.

    e. Click **IP Configuration** to renew the assigned IP address. Click **Static**. Click **DHCP** to receive new IP address information from the wireless router.

    f. Select **Command Prompt**. Enter **ipconfig**.

    Question:

    Record the IP address for PC0:

Part 5: Enable DHCP on the Other PCs

    a. Click **PC1**.

    b. Select the **Desktop** tab.

    c. Select **IP Configuration**.

    d. Click **DHCP**.

    Question:

    Record the IP address for PC1:

    e. Close the configuration window.

    f. Enable DHCP on **PC2** following the steps for PC1.

Part 6: Verify Connectivity

a. Click **PC2** and select the **Desktop** tab.

b. Select **Command Prompt**.

c. Enter **ipconfig** at the prompt to view the IP configuration.

d. At the prompt, enter **ping 192.168.5.1** to ping the wireless router.

e. Enter **ping 192.168.5.126** to ping PC0 at the prompt.

f. At the prompt, enter **ping 192.168.5.127** to ping PC1.

g. The pings to all devices should be successful.

CHAPTER 10

IPv4 and IPv6 Address Management

10.2.3 Packet Tracer–Examine NAT on a Wireless Router

Objectives

- Examine NAT configuration on a wireless router.
- Set up 4 PCs to connect to a wireless router using DHCP.
- Examine traffic that crosses the network using NAT.

Instructions

Part 1: Examine the Configuration for Accessing External Network

a. Add 1 PC and connect it to the wireless router with a straight-through cable. Wait for all link lights to turn green before moving onto the next step or click **Fast Forward**.

b. On the PC, click **Desktop**. Select **IP Configuration**. Click **DHCP** to enable each device to receive an IP address via the DHCP on the wireless router.

c. Note the IP address of the default gateway. Close the **IP Configuration** window when done.

d. Navigate to the web browser and enter the IP address of the default gateway in the URL field. Enter the username **admin** and password **admin** when prompted.

e. Click the **Status** menu option in the upper right-hand corner. When selected, it displays the Router sub-menu page.

f. Scroll down the router page to the Internet connection option. The IP address assigned here is the address assigned by the ISP. If no IP address is present (0.0.0.0 appears), close the window, wait for a few seconds and try again. The wireless router is in the process of obtaining an IP address from the ISP DHCP server.

The address seen here is the address assigned to the Internet port on the wireless router.

Question:

Is this a private or public address?

_____

Part 2: Examine the Configurations for Accessing the Internal Network

a. Click **Local Network** within the Status sub-menu bar.

b. Scroll down to examine the Local Network information. This is the address assigned to the internal network.

c. Scroll down further to examine the DHCP server information, and range of IP addresses that can be assigned to connected hosts.

Question:

Are these private or public addresses?

_____

d. Close the wireless router configuration window.

Part 3: Connect 3 PCs to the Wireless Router

a. Add 3 more PCs and connect them to the wireless router with straight-through cables. Wait for all link lights to turn green before moving on to the next step or click **Fast Forward.**

b. On each PC, click **Desktop.** Select **IP Configuration.** Click **DHCP** to enable each device to receive an IP address via the DHCP on the wireless router. Close the **IP Configuration** window when done.

c. Click **Command Prompt** to verify each device IP configuration using the **ipconfig /all** command.

Note: These devices will receive a private address. Private addresses are not able to cross the Internet; therefore, NAT translation must occur.

Part 4: View NAT Translation Across the Wireless Router

a. Enter **Simulation Mode** by clicking the **Simulation** tab in the lower right-hand corner. The Simulation tab is located next to the Realtime tab and has a stopwatch symbol.

b. View traffic by creating a Complex PDU in **Simulation Mode:**

1) From the **Simulation Panel,** click **Show All/None** to change visible events to none. Now click **Edit Filters** and under the **Misc** tab, checkmark the boxes for **TCP** and **HTTP.** Close the window when done.

2) Add a Complex PDU by clicking on the opened envelope located in upper menu.

3) Click one of the PCs to specify it as the source.

c. Specify the Complex PDU settings by changing the following within the complex PDU window:

1) Under **PDU Settings > Select Application** should be set to: **HTTP.**

2) Click **ciscolearn.nat.com** server to specify it as the destination device.

3) For the **Source Port,** enter **1000.**

    4) Under **Simulation Settings**, select **Periodic**. Enter **120** seconds for the Interval.

    5) Click **Create PDU** in the **Create Complex PDU** window.

  d. Double-click the simulation panel to unlock it from the PT window. This enables you to move the simulation panel to view the entire network topology.

  e. Observe the traffic flow by clicking **Play** in the simulation panel. Speed up the animation by sliding the play control slider to the right.

Note: Click **View Previous Events** when the Buffer Full message is displayed.

Part 5: View the Header Information of the Packets that Traveled Across the Network

  a. Examine the headers of the packets sent between a PC and the web server.

    1) In the **Simulation Panel**, double click the 3rd line down in the event list. This displays an envelope in the work area that represents that line.

    2) Click the envelope in the work area window to view the packet and header information.

  b. Click the **Inbound PDU details** tab. Examine the packet information for the source (SRC) IP address and destination IP address.

  c. Click the **Outbound PDU details** tab. Examine the packet information for the source (SRC) IP address and destination IP address.

    Notice the change in SRC IP address.

  d. Click through other event lines to view those headers throughout the process.

  e. When finished, click **Check Results** to check your work.

10.4.7 Lab–Identify IPv6 Addresses

Topology

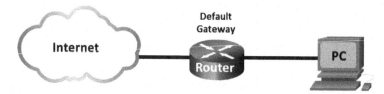

Objectives

- Part 1: Practice with Different Types of IPv6 Addresses.
- Part 2: Examine a Host IPv6 Network Interface and Address.

Background / Scenario

With the depletion of the Internet Protocol version 4 (IPv4) network address space and the adoption and transition to IPv6, networking professionals must understand how both IPv4 and IPv6 networks function. Many devices and applications already support IPv6. This includes extensive Cisco device Internetwork Operating System (IOS) support and workstation/server operating system support, such as that found in Windows and Linux.

This lab focuses on IPv6 addresses and the components of the address. In Part 1, you will identify the IPv6 address types and IPv6 addresses abbreviation. In Part 2, you will view the IPv6 settings on a PC.

Required Resources

- 1 PC (Windows with Internet access)

Instructions

Part 1: Practice with Different Types of IPv6 Addresses

In this part, you will identify the different types of IPv6 addresses and practice compressing and decompressing IPv6 addresses.

Step 1:   Match the IPv6 address to its type.

Match the IPv6 addresses to their corresponding address type. Notice that the addresses have been compressed to their abbreviated notation and that the slash network prefix number is not shown. Some answer choices must be used more than once.

Answer choices:

a.   Loopback address

b.   Global unicast address

    c. Link-local address

    d. Unique-local address

    e. Multicast address

IPv6 Address	Answer
2001:0db8:1:acad::fe55:6789:b210	
::1	
fc00:22:a:2::cd4:23e4:76fa	
2033:db8:1:1:22:a33d:259a:21fe	
fe80::3201:cc01:65b1	
ff00::	
ff00::db7:4322:a231:67c	
ff02::2	

Step 2: Practice compressing and decompressing IPv6 addresses.

Questions:

Using the rules of IPv6 address abbreviation, either compress or decompress the following addresses:

2002:0ec0:0200:0001:0000:04eb:44ce:08a2

fe80:0000:0000:0001:0000:60bb:008e:7402

fe80::7042:b3d7:3dec:84b8

ff00::

2001:0030:0001:acad:0000:330e:10c2:32bf

Part 2: Examine a Host IPv6 Network Interface and Address

In Part 2, you will check the IPv6 network settings of your PC to identify your network interface IPv6 address.

Step 1: Check your PC IPv6 network address settings.

Verify that the IPv6 protocol is installed and active on your PC-A (check your Local Area Connection settings).

    a. Navigate to the **Control Panel**.

    b. In the Category View, click **Network and Sharing Center** icon. Click **View network status and tasks**.

c. In the **Network and Sharing Center** window, you will see your active networks.

d. On the left side of the window, click **Change adapter settings**. You should now see icons representing your installed network adapters. Right-click your active network interface (it may be an **Ethernet** or a **Wi-Fi**), and then click **Properties**.

e. In the Properties window, scroll through the list of items to determine whether IPv6 is present, which indicates that it is installed, and if it is also checkmarked, which indicates that it is active.

f. Select the item **Internet Protocol Version 6 (TCP/IPv6)** and click **Properties**. You should see the IPv6 settings for your network interface. Your IPv6 properties window is likely set to **Obtain an IPv6 address automatically**. This does not mean that IPv6 relies on the Dynamic Host Configuration Protocol (DHCP). Instead of using DHCP, IPv6 looks to the local router for IPv6 network information and then auto-configures its own IPv6 addresses. To manually configure IPv6, you must provide the IPv6 address, the subnet prefix length, and the default gateway. Click **Cancel** to exit the **Properties** windows.

Note: The local router can refer host requests for IPv6 information, especially Domain Name System (DNS) information, to a DHCPv6 server on the network.

---

e. After you have verified that IPv6 is installed and active on your PC, you should check your IPv6 address information.

Step 2: Verify the IPv6 address settings using the Command Prompt.

a. Open a command prompt and enter the **ipconfig /all** command. Your output should look similar to this:

```
C:\Users\user> ipconfig /all

Windows IP Configuration

<output omitted>

Wireless LAN adapter Wireless Network Connection:

    Connection-specific DNS Suffix  . :
    Description . . . . . . . . . . . : Intel(R) Centrino(R) Advanced-N
                                        6200 AGN
    Physical Address. . . . . . . . . : 02-37-10-41-FB-48
    DHCP Enabled. . . . . . . . . . . : Yes
    Autoconfiguration Enabled . . . . : Yes
    Link-local IPv6 Address . . . . . : fe80::8d4f:4f4d:3237:95e2%14
                                        (Preferred)
    IPv4 Address. . . . . . . . . . . : 192.168.2.106(Preferred)
    Subnet Mask . . . . . . . . . . . : 255.255.255.0
    Lease Obtained. . . . . . . . . . : Sunday, January 06, 2013
                                        9:47:36 AM
    Lease Expires . . . . . . . . . . : Monday, January 07, 2013
                                        9:47:38 AM
    Default Gateway . . . . . . . . . : 192.168.2.1
    DHCP Server . . . . . . . . . . . : 192.168.2.1
    DHCPv6 IAID . . . . . . . . . . . : 335554320
```

```
DHCPv6 Client DUID. . . . . . . . : 00-01-00-01-14-57-84-B1-1C-
                                    C1-DE-91-C3-5D
DNS Servers . . . . . . . . . . . : 192.168.1.1
                                    8.8.4.4
<output omitted>
```

b. You can see from the output that the client PC has an IPv6 link-local address with a randomly generated interface ID.

Questions:

What does it indicate about the network regarding IPv6 global unicast address, IPv6 unique-local address, or IPv6 gateway address?

What kind of IPv6 addresses did you find when using **ipconfig /all**?

Reflection Questions

1. How do you think you must support IPv6 in the future?

2. Do you think IPv4 networks continue on, or will everyone eventually switch over to IPv6? How long do you think it will take?

11.1.6 Packet Tracer–The Client Interaction

Objectives

- Observe the client interaction between the server and PC.

Background / Scenario

Clients, such as desktop PCs, request services from servers. The lab environment, using physical PCs and servers, supports a full range of services. In a simulated environment, the number of services is limited. Packet Tracer allows the addition of simulated network servers that support DHCP, DNS, HTTP, and TFTP. Packet Tracer also supports the addition of simulated PCs that can request these services. This activity uses a simple network consisting of a PC connected directly to a server configured to supply DNS services as well as hosting a web page through an HTTP server. This activity will track the flow of traffic that happens when a web page is requested, how the IP address of the web page is resolved, and the web page is delivered.

Instructions

Part 1: Enter Simulation Mode

When Packet Tracer starts, it presents a logical view of the network in realtime mode.

Click the **Simulation Mode** icon to enter Simulation Mode. The Simulation Mode icon is located in the bottom right of the logical workplace.

Part 2: Set Event List Filters

In simulation mode, the default is to capture all events. You will use filters to only capture DNS and HTTP events.

a. In the **Event List Filters** section, click **Show All/None** to clear all the checks.

b. Click **Edit Filters**. Under the IPv4 tab, select **DNS**. Under the **Misc** tab, select **HTTP**. Close the window when done. The **Event List Filters** shows DNS and HTTP as the only visible events.

Part 3: Request a Web Page from the PC

You will open a simulated web browser on the PC and request a web page from the server.

a. Click **PC**. Click **Desktop** tab and click **Web Browser**.

b. A simulated web browser opens. Type **www.example.com** into the URL box and click the **Go** button to the right. Minimize the PC window.

Part 4: Run the Simulation

a. In the **Play Controls** section of the Simulation Panel, click **Play**. The exchange between the PC and the server is animated and the events are added to the **Event List**.

These events represent the PC's request to resolve the URL to an IP address, the server's providing of the IP address, the PC's request for the web page, the server's sending the web page in two segments, and the PC's acknowledging the web page.

b. Click **View Previous Event** to continue when the buffer is full.

Part 5: Access a Specific PDU

a. Restore the simulated PC window. Notice there is a web page displayed in the web browser. Minimize the simulated browser window.

b. In the Simulation Panel **Event List** section, the last column contains a colored box that provides access to detailed information about an event. Click the colored box in the first row for the first event. The **PDU Information** window opens.

Part 6: Examine the Contents of the PDU Information Window

The first tab in the **PDU Information** window contains information about the inbound and/or outbound PDU as it relates to the OSI model. Click **Next Layer >>** repeatedly to cycle through the inbound and outbound layers and read the description in the box below the layers to get an overview of how the exchange works.

Examine the PDU information for the other events to get an overview of the entire exchange process.

CHAPTER 12

Application Layer Services

 ## 12.2.4 Lab–Observe DNS Name Resolution

Objectives

- Observe the conversion of a URL to an IP address.
- Observe DNS lookup using the nslookup command.

Background / Scenario

Domain Name System (DNS) is invoked when you type a Uniform Resource Locator (URL), such as http://www.cisco.com, into a web browser. The first part of the URL describes which protocol is being used. Common protocols are HTTP (Hypertext Transfer Protocol), HTTPS (Hypertext Transfer Protocol over Secure Socket Layer), and FTP (File Transfer Protocol).

DNS uses the second part of the URL, which in this example is www.cisco.com. DNS translates the domain name (like www.cisco.com) to an IP address to allow the source host to reach the destination host.

Work in pairs to complete this lab.

Required Resources

- 1 PC (Windows 10)
- Internet connectivity

Instructions

Part 1: Observe DNS Conversion

Step 1:   Observe DNS operation.

a.   Open a **Command Prompt** window.

b.   At the prompt, enter **ping cisco.com**. The computer needs to translate cisco.com into an IP address so it knows where to send the Internet Control Message Protocol (ICMP) packets. Ping is a type of ICMP packet.

c. The first line of the output shows cisco.com converted to an IP address by DNS. You should be able to see the effect of DNS even if your school has a firewall that prevents pinging, or if Cisco has prevented people from pinging their web server.

```
C:\Users\Student> ping cisco.com

Pinging cisco.com [72.163.4.185] with 32 bytes of data:
Reply from 72.163.4.185: bytes=32 time=34ms TTL=244
Reply from 72.163.4.185: bytes=32 time=32ms TTL=244
Reply from 72.163.4.185: bytes=32 time=34ms TTL=244
Reply from 72.163.4.185: bytes=32 time=34ms TTL=244

Ping statistics for 72.163.4.185:
    Packets: Sent = 4, Received = 4, Lost = 0 (0% loss),
Approximate round trip times in milli-seconds:
    Minimum = 32ms, Maximum = 34ms, Average = 33ms
```

Questions:

Which IP address is shown on the screen?

_____

Should cisco.com always resolve to the same IP address? Explain.

_____

List a few applications (besides the **ping** command) that need DNS to translate a domain name to an IP address. Perform an internet search as necessary.

_____

Step 2: Verify DNS operation using the nslookup command.

a. At the command prompt, enter the **nslookup** command.

```
C:\Users\Student> nslookup
Default Server:  google-public-dns-a.google.com
Address:  8.8.8.8

>
```

The output above shows that the Default DNS Server was configured to use a Google DNS server.

Question:

What is your Default Server listed as?

_____

b. After issuing the previous **nslookup** command, notice how the prompt changed to a single **>**. This is the prompt for the **nslookup** program. From this prompt, you can enter commands related to DNS.

At the prompt, type **?** to see a list of all the available commands that you can use in **nslookup** mode.

Question:

List three commands that you can use with **nslookup**:

_____

c. At the **nslookup** prompt, type **cisco.com**.

```
> cisco.com
```

Questions:

What is the translated IP address?

_____

Is the IP address an IPv4 address or an IPv6 address?

_____

Is it the same as the IP address shown with the **ping** command?

_____

At the prompt, type the IP address of the Cisco web server that you just found. What is the Name result?

_____

Step 3:   **Identify mail servers using the nslookup command**

a. To identify mail servers using **nslookup**, enter **set type=mx**.

```
> set type=mx
```

b. At the prompt, enter **cisco.com**.

```
> cisco.com
Server:  google-public-dns-a.google.com
Address:  8.8.8.8

Non-authoritative answer:
cisco.com       MX preference = 20, mail exchanger = rcdn-mx-01.cisco.com
cisco.com       MX preference = 30, mail exchanger = aer-mx-01.cisco.com
cisco.com       MX preference = 10, mail exchanger = alln-mx-01.cisco.com
>
```

Question:

What are the names of the Cisco mail servers identified in the **mail exchanger** field?

_____

c. At the prompt, type **exit** to return to the regular command prompt.

d. At the prompt, type **ipconfig /all**.

Question:

Write the IP addresses of all the DNS servers that your school computer uses.

e. Enter **exit** to close the Command Prompt window.

Reflection

1. If your school did not have a DNS server, what effect would this have on your use of the Internet?

2. Some companies do not dedicate a single server for DNS. Instead, the DNS server provides other functions as well. Which functions do you think might be included on a DNS server? Use the **ipconfig /all** command to help you with this.

12.3.3 Packet Tracer–Observe Web Request

Objectives

View the client/server traffic sent from a PC to a web server when requesting web services.

Instructions

Part 1: Verify Connectivity to the Web Server

a. Click **External Client** and access the **Command Prompt** from the **Desktop** tab.

b. Use the **ping** command to reach the URL **ciscolearn.web.com**.

    PC> ping ciscolearn.web.com

Notice the IP address included in the **ping** output. This address is obtained from the DNS server and resolves to the domain name ciscolearn.web.com. All traffic forwarded across a network uses source and destination IP address information.

c. Close the **Command Prompt** window but leave the **External Client** Desktop window open.

Part 2: Connect to the Web Server

a. From the **Desktop** window, access the **Web Browser**.

b. In the URL block, type **ciscolearn.web.com**.

*Be sure to read the web page that is displayed. Leave this page open.

c. Minimize the **External Client** window but do not close it.

Part 3: View the HTML Code

a. From the Logical topology, click **ciscolearn.web.com** server.

b. Click the **Services** tab > **HTTP** tab. Then next to the **index.html** file click **(edit)**.

c. Compare the HTML markup code on the server that creates the Web Browser display page on the **External Client**. This may require that you re-maximize the **External Client** window if it shrunk when you opened the server window.

d. Close both the **External Client** and web server windows.

Part 4: Observe Traffic Between the Client and the Web Server

a. Enter Simulation mode by clicking the **Simulation** tab in the lower right-hand corner.

b. Double click the Simulation Panel to unlock it from the PT window. This allows you to move the Simulation Panel to view the entire network topology.

c. View traffic by creating a Complex PDU in Simulation Mode.

    1) From the **Simulation Panel,** select **Edit Filters.**

    2) Click the **Misc** tab to verify that only the boxes for TCP and HTTP are selected.

    3) Add a Complex PDU by clicking the open envelope located above the Simulation Mode icon.

    4) Click the **External Client** to specify it as the source. The **Create Complex PDU** window will appear.

d. Specify the **Create Complex PDU** settings by changing the following within the Complex PDU window:

    1) Under PDU Settings, **Select Application** should be set to **HTTP.**

    2) Click the **ciscolearn.web.com** server to specify it as the destination device. Notice the IP address of the web server will appear in the destination box within the complex PDU window

    3) For the **Starting Source Port,** enter **1000.**

    4) Under **Simulation Settings,** select **Periodic Interval** and type **120** seconds.

e. Create the PDU by clicking the box **Create PDU** in the **Create Complex PDU** window.

    1) Observe the traffic flow by clicking **Play** in the **Simulation Panel.** Speed up the animation by using the play control slider.

    When the **Buffer Full** window appears, click **View Previous Events** to close the window.

    2) Scroll through the Event List. Notice the number of packets that traveled from source to destination. HTTP is a TCP protocol, which requires connection establishment and acknowledgment of receipt of packets, considerably increasing the amount of traffic overhead.

Packet Tracer
□ Activity

12.4.4 Packet Tracer–Use FTP Services

Addressing Table

Device	Interface	IP Address	Subnet Mask
FTP Server (ftp.pka)	NIC	209.165.200.226	255.255.255.224

Objectives

- Upload a file to an FTP server.
- Download a file from an FTP server.

Background / Scenario

File Transfer Protocol (FTP) is a commonly used application to transfer files between clients and servers on the network. The server is configured to run the service where clients connect, login, and transfer files. FTP uses port 21 as the server command port to create the connection. FTP then uses port 20 for data transfer.

In this activity, you will upload a file to an FTP server. You will also download a file from an FTP server.

Instructions

Part 1: Upload a File to an FTP Server

In this part, you will locate the file **sampleFile.txt** and upload it to an FTP server.

Step 1:   Locate the file.

   a.   Click **PC-A**.

   b.   Click **Desktop**.

   c.   Click **Command Prompt**.

   d.   At the prompt, click **?** to list the available commands.

   e.   Enter **dir** to see the files on the PC. Notice that there is a **sampleFile.txt** file in the C:\ directory.

```
C:\> dir
Volume in drive C has no label.
Volume Serial Number is 5E12-4AF3
Directory of C:\

12/31/1969 17:0 PM 26 sampleFile.txt
26 bytes 1 File(s)
```

Step 2:   Connect to the FTP server.

   a.   FTP to the FTP server at **209.165.200.226** or **ftp.pka**.

```
C:\> ftp 209.165.200.226
Trying to connect...209.165.200.226
Connected to 209.165.200.226
```

b. Enter the username **student** and password **class** to gain access.

```
220- Welcome to PT Ftp server
Username:student
331- Username ok, need password
Password:
230- Logged in
(passive mode On)
```

Step 3: Upload a file to an FTP server.

a. Enter **?** to see the commands available in the ftp client.

```
ftp> ?
                ?
                cd
                delete
                dir
                get
                help
                passive
                put
                pwd
                quit
                rename
ftp>
```

b. Enter **dir** to see the files available on the server.

```
ftp> dir

Listing /ftp directory from 192.168.1.3:
0 : asa842-k8.bin 5571584
1 : asa923-k8.bin 30468096
2 : c1841-advipservicesk9-mz.124-15.T1.bin 33591768
3 : c1841-ipbase-mz.123-14.T7.bin 13832032
<output omitted>
```

c. Enter **put sampleFile.txt** to send the file to the server.

```
ftp> put sampleFile.txt

Writing file sampleFile.txt to 209.165.200.226:
File transfer in progress...

[Transfer complete - 26 bytes]

26 bytes copied in 0.08 secs (325 bytes/sec)
ftp>
```

d. Use the **dir** command again to list the contents of the FTP server to verify that the file has been uploaded to the FTP server.

Part 2: Download a File from an FTP Server

You can also download a file from an FTP server. In this part, you will rename the file sampleFile.txt and download it from the FTP server.

Step 1:   Rename the file on an FTP server.

    a. At the **ftp>** prompt, rename the file **sampleFile.txt** to **sampleFile_FTP.txt**.

```
ftp> rename sampleFile.txt sampleFile_FTP.txt

Renaming sampleFile.txt
ftp>
[OK Renamed file successfully from sampleFile.txt to sampleFile_FTP.txt]
ftp>
```

    b. At the **ftp>** prompt, enter **dir** to verify the file has been renamed.

Step 2:   Download the file from the FTP server.

    a. Enter the command **get sampleFile_FTP.txt** to retrieve the file from the server.

```
ftp> get sampleFile_FTP.txt

Reading file sampleFile_FTP.txt from 209.165.200.226:
File transfer in progress...

[Transfer complete - 26 bytes]

26 bytes copied in 0.013 secs (2000 bytes/sec)
ftp>
```

    b. Enter **quit** to exit the FTP client when finished.

    c. Display the contents of the directory on the PC again to see the image file from the FTP server.

Step 3:   Delete the file from the FTP server.

    a. Log into the FTP server again to delete the file **sampleFile_FTP.txt**.

    b. Enter the command to delete the file **sampleFile_FTP.txt** from the server.

    Question:

    What command did you use to remove the file from the FTP server?

    _____

    c. Enter **quit** to exit the FTP client when finished.

12.5.4 Packet Tracer–Use Telnet and SSH

Addressing Table

Device	Interface	IP Address	Subnet Mask
HQ	G0/0/1	64.100.1.1	255.255.255.0
PC0	NIC	DHCP	
PC1	NIC	DHCP	

Objectives

In this activity, you will establish a remote connection to a router using Telnet and SSH.

- Verify connectivity
- Access a remote device

Instructions

Part 1: Verify Connectivity

In this part, you will verify that the PC has IP addressing and can ping the remote router.

Step 1:   Verify IP address on a PC.

    a.   From a PC, click **Desktop**. Click **Command Prompt**.

    b.   At the prompt, verify that the PC has an IP address from DHCP.

      Question:

      What command did you use to verify the IP address from DHCP?

Step 2:   Verify connectivity to HQ.

    Verify that you can ping the router **HQ** using the IP address listed in the Addressing Table.

Part 2: Access a Remote Device

In this part, you will attempt to establish a remote connection using Telnet and SSH.

Step 1:   Telnet to HQ.

    At the prompt, enter the command **telnet 64.100.1.1.**

    Question:

    Were you successful? What was the output?

Step 2:   SSH to HQ.

The router is properly configured to not allow insecure Telnet access. You must use SSH. **HQ** is not configured to accept Telnet traffic. You cannot connect to the router via the Telnet protocol. You will attempt to use SSH to connect to **HQ**.

The router is configured with a local username **admin** with the password **class** for SSH access.

At the prompt, enter the command **ssh -l admin 64.100.1.1**. Enter the password **class** when prompted.

```
C:\> ssh -l admin 64.100.1.1

Password:
```

Question:

What is the prompt after accessing the router successfully via SSH?

13.5.5 Lab–Configure a Wireless Router and Client

Topology

Wireless Router

Sample Wireless Router Settings

Network Name (SSID)	Network Passphrase	Router Password
Home-Net	cisco123	cisco12345

Note: The above wireless router settings are used as an example only.

Objectives

Part 1: Configure Basic Settings on a Wireless Router.

Part 2: Connect a Wireless Client.

Background / Scenario

It is common to access the Internet from anywhere in the home or office today. Wireless connectivity is what makes that possible. Users have embraced the flexibility that wireless routers provide for accessing the network and the Internet.

In this lab, you will configure a wireless router, which includes applying WPA2 security settings and activating DHCP services. You will also configure a wireless PC client.

Required Resources

- 1 PC with a wireless NIC (Windows 10)
- 1 wireless router
- Ethernet cables as shown in the topology

Instructions

Part 1: Configure Basic Settings on a Wireless Router

One way to configure basic settings on a wireless router is to run the setup CD that came with the router. If you are using the setup CD, you can usually follow the directions provides on the setup CD. If the setup CD is unavailable, download the setup program from the wireless router manufacturer.

If you choose to manually set up the wireless router, the following directions can be used as a guideline to set up the router.

Step 1:   Cable the network as shown in the topology.

    a.   Connect a PC to an unused Ethernet port on the wireless router.

    b.   Connect a power cable to the wireless router. Power up the wireless router. Allow time for the router to boot up.

Step 2:   Configure the wireless settings.

    a.   From the web browser on the PC, enter the IP address to connect to the wireless router. Most wireless routers use http://192.168.1.1, http://192.168.0.1, or http://192.168.2.1. Enter the default username and password as provided by the manufacturer if necessary to log into the web interface.

    b.   For the Internet setup, use DHCP for the Internet IP address unless the ISP provided you with a static IP address.

    c.   For the local network, enable the DHCP server and use 192.168.100.1/24 as the internal network. The starting IP address is 192.168.100.100, and this network allows a maximum number of 150 DHCP users.

> **Note:** The subnet 192.168.100.0/24 is only used as an example of the lab.

    d.   Renew the IP address on the PC to continue. Enter **192.168.100.1** in the web browser to access the wireless router web interface.

    e.   Configure the SSID as **Home-Net**.

    f.   Configure the wireless security by setting the authentication type to use **WPA2 Personal** and set the passphrase to **cisco123**.

    g.   Change the administrative password from the default password to **cisco12345**.

Part 2: Connect a Wireless Client

In Part 2, you will configure the PC's wireless NIC to connect to the wireless router.

> **Note:** This lab was performed using a PC running Windows 10 operating system. You should be able to perform the lab with other Windows operating systems listed. However, menu selections and screens may vary.

Step 1:   Use the Network and Sharing Center.

    a.   Open the **Network Connections** window by right-clicking **Start** > select **Network Connections**.

b. The **Network Connections** window displays the list of NICs available on this PC. Look for your Local Area Connection and Wireless Network Connection adapters in this window.

Note: Other types of network adapters, such as Bluetooth Network connection and Virtual Private Network (VPN) adapter may also be displayed in this window.

Step 2: Work with your wireless NIC.

a. Right-click **Wireless Network Connection** to see the available options. The first option displays if your wireless NIC is enabled. Currently, this NIC is enabled because the **Disable** option is displayed. If your wireless NIC is disabled, you will have an option to **Enable** it.

b. Click **Connect/Disconnect** to open the **Network & Internet** Settings window.

c. Click on **Home-Net** and select **Disconnect**.

d. A list of SSIDs in range of your wireless NIC is displayed. Re-connect with Home-Net. Select Home-Net, and click **Connect**.

e. If prompted, enter **cisco123** to supply the network security key, and then click **OK**.

Step 3: Verify internet access.

a. Verify that the PC received an IP address from the router via DHCP.

b. Open a web browser and navigate to a website, such as www.cisco.com.

14.2.13 Lab–Install Linux in a Virtual Machine and Explore the GUI

Objectives

Part 1: Prepare a Computer for Virtualization.

Part 2: Install a Linux OS on the Virtual Machine.

Part 3: Explore the GUI.

Background / Scenario

Computing power and resources have increased tremendously over the last ten years. A benefit of multi-core processors and large amounts of RAM is the capability to install multiple operating systems through the use of virtualization on a computer.

With virtualization, one or more virtual computers can operate inside one physical computer. Virtual computers that run within physical computers are called virtual machines. Virtual machines are often called guests, and physical computers are often called hosts. Anyone with a modern computer and operating system can run virtual machines.

In this lab, you will install a Linux OS in a virtual machine using a desktop virtualization application, such as VirtualBox. After completing the installation, you will explore the GUI interface. You will also explore the command line interface using this virtual machine in a lab later in this course.

Required Resources

- Computer with a minimum of 2 GB of RAM and 10 GB of free disk space
- High-speed Internet access to download Oracle VirtualBox and Linux OS image, such as Ubuntu Desktop

Instructions

Part 1: Prepare a Computer for Virtualization

In Part 1, you will download and install desktop virtualization software and a Linux OS image. Your instructor may provide you with a Linux OS image.

Step 1:   Download and install VirtualBox.

VMware Player and Oracle VirtualBox are two virtualization programs that you can download and install to support the OS image file. In this lab, you will use the VirtualBox application.

a.   Navigate to https://www.virtualbox.org/. Click the download link on this page.

b.   Choose and download the appropriate installation file based on your operating system.

c.   After the VirtualBox installation file is downloaded, run the installer and accept the default installation settings.

Step 2:   Download a Linux Image.

a.   Navigate to the Ubuntu website at http://www.ubuntu.com. Click the **Download** link on this page to download and save an Ubuntu Desktop image.

Step 3:   Create a New Virtual Machine.

a.   Click **Start** and search for **Virtualbox**. Click **Oracle VM VirtualBox** to open the manager. When the manager opens, click **New** to start the Ubuntu installation.

b.   In the **Name and operating system** screen, type **Ubuntu** in the **Name** field. For the **Type** field, select **Linux**. In the **Version** field, select the corresponding downloaded version. Click **Next** to continue.

c.   In the **Memory size** screen, increase the amount of RAM as long as the amount of RAM for the virtual machine is in the green area. Going beyond the green area would adversely affect the performance of the host. Click **Next** to continue.

d.   In the **Hard disk** screen, click **Create** to create a virtual hard disk now.

e.   In the **Hard disk file type** screen, use the default file type settings of **VDI (VirtualBox Disk Image)**. Click **Next** to continue.

f.   In the **Storage on physical hard disk** screen, use the default storage settings of **dynamically allocated**. Click **Next** to continue.

g.   In the **File location and size** screen, you can adjust the hard drive and change the name and location of the virtual hard drive. Click **Create** to use the default settings.

h.   When the hard drive creation is done, the new virtual machine is listed in the **Oracle VM VirtualBox Manager** window. Select **Ubuntu** and click **Start** in the top menu.

Part 2: Install Ubuntu on the Virtual Machine

Step 1:   Mount the Image.

a.   In the Oracle **VM VirtualBox Manager** window, right-click **Ubuntu** and select **Settings**. In the **Ubuntu – Settings** window, click **Storage** in the left pane. Click **Empty** in the middle pane. In the right pane, click the CD symbol and select the file location of the Ubuntu image. Click **OK** to continue.

b. In the **Oracle VM VirtualBox Manager** window, click **Start** in the top menu.

Step 2:   Install the OS.

a. In the **Welcome** screen, you are prompted to try or install Ubuntu. The try option does not install the OS, it runs the OS straight from the image. In this lab, you will install the Ubuntu OS in this virtual machine. Click **Install Ubuntu.**

b. Follow the on-screen instructions and provide the necessary information when prompted.

Note: If you are not connected to the Internet, you can continue to install and enable the network later.

---

c. Because this Ubuntu installation is in a virtual machine, it is safe to erase the disk and install Ubuntu without affecting the host computer. Select **Erase disk and install Ubuntu.** Otherwise installing Ubuntu on a physical computer would erase all data on the disk and replace the existing operating system with Ubuntu. Click **Install Now** to start the installation.

d. Click **Continue** to erase the disk and install Ubuntu.

e In the **Who are you?** screen, provide your name and choose a password. You can use the username generated or enter a different username. Enter your desired username and password. If desired, you can change the other settings. Click **Continue.**

f. The Ubuntu OS is now installing in the virtual machine. This will take several minutes. When the **Installation is complete** message displays, return to the **Oracle VM Virtualbox Manager** window. Right-click **Ubuntu** and select **Settings**. In the **Ubuntu – Settings** window, click **Storage** in the left pane. Click the mounted Ubuntu image in the middle pane. In the right pane, click the CD symbol and click **Remove Disk from Virtual Drive**. Click **OK** to continue.

g. In the Ubuntu VM, click **Restart Now.**

Part 3: Explore the GUI

In this part, you will install the VirtualBox guest additions and explore the Ubuntu GUI.

Step 1:   Install Guest Additions.

    a. Log on to your Ubuntu virtual machine using the user credentials created in the previous part.

    b. Your Ubuntu Desktop window may be smaller than expected. This is especially true on high-resolution displays. Click **Device > Insert Guest Additions CD image...** to install the Guest Additions. This allows more functions, such as changing the screen resolution in the virtual machine.

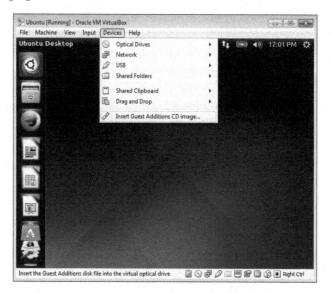

    c. Click **Run** to install the additions. When prompted for a password, use the same password that you used to log on. Click **Authenticate** to continue.

    d. If the computer was not connected to the Internet during the installation, click **Devices > Network Settings** in the Oracle VirtualBox menu. Enable network adapters and configure the proper setting for network connections as necessary. Click **OK**.

    e. When the installation of the additions is done, restart the virtual machine again. Click the menu in the upper-right corner and click **Shut down**. Click **Restart** to restart Ubuntu.

Step 2:   Open a web browser.

    a. Log on to Ubuntu again. After you are logged on again, you can resize the virtual machine window.

    b. Open a web browser. Depending on the Linux distribution, you may need to search for a web browser or there **may be** a link to a web browser already on the desktop.

    c. Locate a terminal emulator to access the command line interface. You will be using a terminal emulator in later labs.

    d. Explore the installed Linux distribution and locate a few applications that you may use.

Reflection Question

What are the advantages and disadvantages of using a virtual machine?

14.3.8 Lab–Configure a Mobile Device for Wi-Fi Connectivity

Objectives

- Familiarize yourself with the Wi-Fi settings on mobile devices.
- Turn the Wi-Fi radio on and off.
- Set the device to forget a found Wi-Fi network.
- Connect to a new Wi-Fi network.

Background / Scenario

In this lab, you will access smartphone operating systems to enable Wi-Fi communications and configure connection settings.

Required Resources

- Android tablet or smartphone running Android version 4.0 or above
- iOS tablet or smartphone running iOS version 5.0 or above
- Access to a configured wireless network, either within the classroom or a wireless hotspot

Instructions

Part 1: Configure Wi-Fi Settings on an Android Device

Step 1: Access the device.

Turn on the device and log in with the password, pin code, or other passcode, if necessary. The main home screen appears.

Step 2: Forget a wireless network.

a. Touch the Settings icon.

b. In the Settings menu, touch the Wi-Fi switch until it is set to Off.

c. Touch the Wi-Fi switch again.

d. Touch the name of the network to which the device is connected.

e. In the Wi-Fi window, touch Forget.

Question:

What happens?

Step 3: Connect to a wireless network.

a. Touch the network to which the device used to be connected.

b. Enter the Wi-Fi Password. Touch Done.

c. Touch Connect.

Question:

What happens?

Part 2: Configure Wi-Fi Settings on an iOS Device

Step 1:   Access the device.

Turn on the device and log in with the password, pin code, or other passcode, if necessary. The main home screen appears.

Step 2:   Forget a wireless network.

    a.   From the Home screen, touch **Settings**.

    b.   Touch **Wi-Fi**.

    c.   Slide the Wi-Fi switch until it is **OFF**.

    c.   Slide the Wi-Fi switch until it is **ON**.

    d.   Touch the name of the network to which the device is connected.

    e.   Touch **Forget**.

Question:

What happens?

Step 3:   Connect to a wireless network.

    a.   Touch the network to which the device used to be connected.

    b.   Enter the Wi-Fi Password. Touch **Join**.

Question:

What happens?

CHAPTER 15

Security Considerations

 ## 15.2.3 Lab - Social Engineering

Objective

- In this lab, you will research examples of social engineering and identify ways to recognize and prevent it.

Resources

- Computer with Internet access

Instructions

Part 1: Research Social Engineering Examples

Social engineering, as it relates to information security, is used to describe the techniques used by a person (or persons) who manipulate people to access or compromise information about an organization or its computer systems. A social engineer is usually difficult to identify and may claim to be a new employee, a repair person, or researcher. The social engineer might even offer credentials to support that identity. By gaining trust and asking questions, he or she may be able to piece together enough information to infiltrate an organization's network.

Question:

Use any internet browser to research incidents of social engineering. Summarize three examples found in your research.

Part 2: Recognize the Signs of Social Engineering

Social engineers are nothing more than thieves and spies. Instead of hacking their way into your network via the Internet, they attempt to gain access by relying on a person's desire to be accommodating. Although not specific to network security, the scenario below, described in Christopher Hadnagy's book, *The Art of Human Hacking*, illustrates how an unsuspecting person can unwittingly give away confidential information.

> *"The cafe was relatively quiet as I, dressed in a suit, sat at an empty table. I placed my briefcase on the table and waited for a suitable victim. Soon, just such a victim arrived with a friend and sat at the table next to mine. She placed her bag on the seat beside her, pulling the seat close and keeping her hand on the bag at all times.*
>
> *After a few minutes, her friend left to find a restroom. The mark [target] was alone, so I gave Alex and Jess the signal. Playing a couple, Alex and Jess asked the mark if she would take a picture of them both. She was happy to do so. She removed her hand from her bag to take the camera and snap a picture of the "happy couple" and, while distracted, I reached over, took her bag, and locked it inside my briefcase. My victim had yet to notice her purse was missing as Alex and Jess left the café. Alex then went to a nearby parking garage.*

It didn't take long for her to realize her bag was gone. She began to panic, looking around frantically. This was exactly what we were hoping for so, I asked her if she needed help.

She asked me if I had seen anything. I told her I hadn't but convinced her to sit down and think about what was in the bag. A phone. Make-up. A little cash. And her credit cards. Bingo!

I asked who she banked with and then told her that I worked for that bank. What a stroke of luck! I reassured her that everything would be fine, but she would need to cancel her credit card right away. I called the "help-desk" number, which was actually Alex, and handed my phone to her.

Alex was in a van in the parking garage. On the dashboard, a CD player was playing office noises. He assured the mark that her card could easily be canceled but, to verify her identity, she needed to enter her PIN on the keypad of the phone she was using. My phone and my keypad.

When we had her PIN, I left. If we were real thieves, we would have had access to her account via ATM withdrawals and PIN purchases. Fortunately for her, it was just a TV show."

Remember: "Those who build walls think differently than those who seek to go over, under, around, or through them." Paul Wilson - The Real Hustle

Question:

Research ways to recognize social engineering. Describe three examples found in your research.

Part 3: Research Ways to Prevent Social Engineering

Questions:

Does your company or school have procedures in place to help prevent social engineering?

If so, what are some of those procedures?

Use the internet to research procedures that other organizations use to prevent social engineers from gaining access to confidential information. List your findings.

CHAPTER 16

Configure Network and Device Security

16.2.4 Packet Tracer–Configure Basic Wireless Security

Packet Tracer □ Activity

Objectives

- Configure basic wireless security using WPA2 Personal.

Background / Scenario

A small business owner learns that the wireless network should be secured from unauthorized access. He has decided to use WPA2 Personal for his network.

Instructions

Part 1: Verify Connectivity

a. Access the **Desktop > Web Browser** on the laptop.

b. Enter **www.cisco.pka** into the URL field. The web page should display.

Part 2: Configure Basic Wireless Security

a. Enter **192.168.1.1** in the web browser to access the wireless router. Enter **admin** as the username and password.

b. Click the **Wireless** menu. Select the **Wireless Security** menu.

c. The security mode is disabled currently. For a 2.4 GHz network, change the security mode to **WPA2 Personal**. For a 5 GHz network, you can leave it disabled.

d. In the Passphrase field, enter **Network123**.

e. Scroll down to the bottom of the page and click **Save Settings**. Close the web browser.

Part 3: Update the Wireless Settings on the Laptop

a. Click **PC Wireless** in the **Desktop** tab.

b. Click the **Connect** tab. Select the **Academy** and click **Connect**.

c. Enter **Network123** as the pre-shared key. Click **Connect**.

d. Close the **PC Wireless** window.

Part 4: Verify Connectivity

a. Access the web browser.

b. Enter **www.cisco.pka** into the URL field. Verify that the web page displays after the addition of basic wireless configuration.

c. If you are unable to access the web page, verify your wireless settings on the wireless router and the laptop. Also verify that the laptop is connected to the wireless router.

16.3.8 Lab–Configure Firewall Settings

Topology

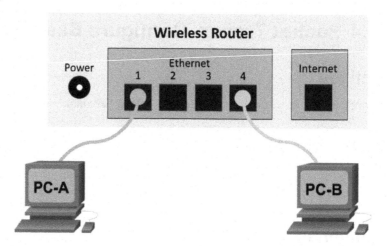

Addressing Table

Device	Interface	IP Address	Subnet Mask
PC-A	NIC	192.168.1.10	255.255.255.0
PC-B	NIC	192.168.1.11	255.255.255.0

Objectives

- Access Windows Firewall settings to add a new firewall rule.
- Create a firewall rule to permit ping requests.
- Remove the new firewall rule to return the settings to their previous state.

Background / Scenario

If the members of your team are unable to ping your PC, the firewall may be blocking those requests. Many of the labs in the course suggest that you disable the Windows firewall to permit ping requests and responses to execute correctly. Disabling a firewall is not a good recommended practice in an actual production network.

In this lab, you will create a rule in the firewall to allow ping requests without putting the PC at risk for other types of attacks. It also describes how to disable the new ICMP rule after you have completed the lab.

Required Resources

- 1 wireless router with at least two switchports
- 2 PCs (Windows 10)
- Two Ethernet cables as shown in the topology

Instructions

Part 1: Verify that the Windows Firewall is Active and is Blocking ICMP Requests

a. Right-click **Start**. Select **Network Connections**.

b. Click **Change adapter options**. Right-click the desired network adapter and select **Properties**.

c. Select **Internet Protocol Version 4 (TCP/IPv4)**. Click **Properties** to configure the two PCs with the static IP addresses shown in the addressing table. No configuration is necessary for a default gateway or a DNS server in this lab, as both PCs are on the same IP network and will use IP addresses instead of domain names.

d. Open a command prompt window on PC-A by right-click **Start > Command Prompt**. Attempt to **ping** the IP address assigned to PC-B. The **ping** command should fail. Repeat the **ping** command on PC-B, attempting to **ping** the address of PC-A. **Ping** commands from both PCs should fail, indicating that the Windows firewall is active and is blocking ICMP ping requests.

Note: If the **ping** succeeds on either PC, verify that the Windows Firewall is active on both machines.

Part 2: Create a New Inbound Rule Allowing ICMP Traffic Through the Firewall

Step 1:   Access **Windows Firewall** advanced settings.

a. Configure the firewall settings on PC-A. Click **Start** and type **Firewall**. Select **Windows Firewall** or **Windows Defender Firewall** from the results list.

b. In the left pane of the Windows Firewall window, click **Advanced settings**.

Step 2:   Create a new inbound rule.

a. On the Advanced Security window, select **Inbound Rules**. Right-click **Inbound Rules** and select **New Rule....**

b. In the **New Inbound Rule** wizard, click **Custom** in the Rule Type screen. Click **Next** to continue.

c. In the left pane, click the **Protocol and Ports** option. In the Protocol type drop-down menu, select **ICMPv4**, and then click **Next**.

Question:

List three protocols, in addition to ICMP, that can be filtered by a new inbound firewall rule.

---

d. In the left pane, click the **Name** option and in the Name field, type **Allow ICMP Requests**. Click **Finish**.

This new rule should allow your team members to receive **ping** replies from PC-A. Repeat the commands in Step 2 to add the new rule on PC-B.

e. Test the new firewall rule by repeating the **ping** command. The pings should be successful.

If not, review the firewall settings to ensure that the new rule is configured correctly.

Part 3: Disabling or Deleting the New ICMP Rule

After the lab is complete, you may want to disable or even delete the new rule you created in Step 2. Using the **Disable Rule** option allows you to enable the rule again. Deleting the rule permanently deletes it from the list of Inbound Rules.

a. On the Advanced Security window, in the left pane, click **Inbound Rules** and then locate the rule you created in a previous step.

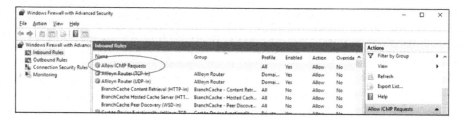

b. To disable the rule, right-click **Inbound Rules** and select **Disable Rule**. When you choose this option, you will see this option change to **Enable Rule**. You can toggle back and forth between Disable Rule and Enable Rule; the status of the rule also shows in the Enabled column of the Inbound Rules list.

c. To permanently delete the ICMP rule, click **Delete**. If you choose this option, you must re-create the rule again to allow ICMP replies.

d. Execute the **ping** commands performed in Step 1 to verify that the firewall is now blocking the ping requests again.

CHAPTER 17

Cisco Switches and Routers

17.5.1 Packet Tracer–Compare In-Band and Out-of-Band Management Access

Addressing Table

Device	Interface	IP Address	Subnet Mask
East	G0/0/0	209.165.200.226	255.255.255.224
	G0/0/1	192.168.1.1	255.255.255.0
West	G0/0/1	64.100.1.1	255.255.255.0
PC	NIC	DHCP	

Objectives

In this activity, you will complete the following objectives.

- Establish an out-of-band connection.

- Establish an in-band connection.

Note: This activity opens in Physical mode. You will see a representation of the devices in an equipment rack. The PC is on a table next to the rack. You can switch to Simulation mode by clicking the button in the blue bar at the top.

Instructions

Part 1: Establish an Out-of-Band Connection

Out-of-band connections are typically used for local access to the device. No IP addressing is required. The network technician has physical access to the device and can connect a cable to either the console or USB port between the device and a PC or laptop. The technician can then use terminal emulation software to connect to the command line interface.

In this part, you will use the console ports to connect to a router and a switch. Then you will use the USB port to connect to another router.

Step 1: Connect the console cable.

    a. Using the device selection box, select the **Console** cable to connect the PC to the console port of the **Router**.

    b. Click the **Router**. Select **Console** in the pop-up menu.

    c. Locate the **PC** on the table. You will need to scroll down and to the right of the Rack to locate it.

    d. Click the **PC**. Select **RS-232** in the pop-up menu. You should now see a blue console cable connected to the correct port on the **Router** and the **PC**.

Step 2:   Access the router.

    a. Click **PC**.

    b. Click the **Desktop** tab. Click **Terminal**.

    c. Review the **Terminal Configuration** settings. These are the correct console settings for accessing a Cisco device using a console connection. Click **OK** to access the **Router**.

    d. At this point, there is no output because the router is powered off. Close the **Terminal**.

Note: With physical equipment, you will not need to reopen the terminal when powering on devices or move the console cable to a different device.

---

    e. To power up the **Router**, click the **Router**. Power up the **Router**. Navigate to the **Physical** tab. Click the **Power** switch. Zoom in as necessary to locate the power switch.

    f. Connect to the **Router** again using the **Terminal** on the PC. You will see the router output. Enter **n** to skip the initial configuration dialog. Press **Enter** to start configuration.

    Question:

    What is the prompt?

    _____

    g. Close the terminal window when done.

    h. Disconnect the console cable from the PC and the router when finished. Click **Delete** on the toolbar at the top and click one end of the console cable to disconnect the console cable. Click **Delete** again to de-select the tool.

Step 3:   Access the switch.

You can also connect to the switch for configuration via the console port.

    a. Connect a console cable from the **Switch** to the **PC** using the same procedure as the previous step.

    b. On the PC, navigate to the **Terminal**. Press **Enter** to see the prompt.

    Question:

    What is the prompt?

    _____

    c. Close the terminal window when done.

    d. Disconnect the console cable from the PC and switch when finished.

Step 4:   Access East via a USB console connection.

For some of the newer routers, you can also connect to the router via the USB console port from the USB port on a PC.

    a. Using the device selection box, select the **USB** cable to connect the PC to the USB console port of the router **East**.

    b. Click **East**. Select **USB Console** in the pop-up menu.

c. Click the **PC**. Select **USB0** in the pop-up menu.

d. Click the **PC**. Click the **Desktop** tab. Click **Terminal**. Click **OK** to access the router **East**.

e. Press **Enter** a few times to see the output.

Part 2: Establish an In-Band Connection

In-band connections are established across an existing network connection between the laptop or PC and the device you wish to access. For this type of connection, an IP address is required. However, you do not need to use terminal emulation software. You can simply use the command line on any device that can access to the remote device using IP addressing. In this part, you will access the routers **East** and **West** via an in-band connection.

The routers **East** and **West** are configured with a local username **admin** with the password **class** for SSH access.

Step 1: Use the existing **East** console connection to establish an in-band connection to the **West** router.

a. At the prompt of the router **East**, enter the command **ssh -l admin 64.100.1.1**. Enter the password **class** when prompted.

```
East> ssh -l admin 64.100.1.1
Password:
```

Question:

What is the prompt after accessing the router successfully via SSH?

_____

b. Close the **Terminal** when finished.

Step 2: From the PC, establish an in-band connection to the East router.

a. Click the **PC**. Then select **Command Prompt**.

b. Enter the **SSH** command to access the router **East**.

Questions:

What command did you use to access the router **East**?

_____

What is the prompt after accessing the router successfully via SSH?

_____

c. At the prompt, enter **exit** to exit the SSH session.

CHAPTER 18

The Cisco IOS Command Line

18.2.6 Packet Tracer–Navigate the IOS

Objectives

- Part 1: Establish Basic Connections, Access the CLI, and Explore Help.
- Part 2: Explore EXEC Modes.
- Part 3: Set the Clock.

Background / Scenario

In this activity, you will practice skills necessary for navigating the Cisco IOS, such as different user access modes, various configuration modes, and common commands used on a regular basis. You will also practice accessing the context-sensitive Help by configuring the **clock** command.

Instructions

Part 1: Establish Basic Connections, Access the CLI, and Explore Help

Step 1: Connect PC1 to S1 using a console cable.

    a. Click the **Connections** icon (the one that looks like a lightning bolt) in the lower left corner of the **Packet Tracer** window.

    b. Select the light blue **Console** cable by clicking it. The mouse pointer will change to what appears to be a connector with a cable dangling from it.

    c. Click **PC1**. A window displays an option for an RS-232 connection. Connect the cable to the RS-232 port.

    d. Drag the other end of the console connection to the S1 switch and click the switch to access the connection list.

    e. Select the **Console** port to complete the connection.

Step 2: Establish a terminal session with S1.

    a. Click **PC1** and then select the **Desktop** tab.

    b. Click the **Terminal** application icon. Verify that the **Port Configuration** default settings are correct.

    Question:

    What is the setting for bits per second?

    _____

    c. Click **OK**.

d. The screen that appears may have several messages displayed. Somewhere on the screen there should be a **Press RETURN to get started!** message. Press **ENTER.**

Question:

What is the prompt displayed on the screen?

_____

Step 3:   Explore the IOS Help.

a. The IOS can provide help for commands depending on the level accessed. The prompt currently displayed is called **User EXEC**, and the device is waiting for a command. The most basic form of help is to type a question mark (?) at the prompt to display a list of commands.

Open Configuration Window

`S1> ?`

Question:

Which command begins with the letter 'C'?

_____

b. At the prompt, type **t** and then a question mark (?).

`S1> t?`

Question:

Which commands are displayed?

_____

At the prompt, type **te** and then a question mark (?).

`S1> te?`

Question:

Which commands are displayed?

_____

This type of help is known as context-sensitive help. It provides more information as the commands are expanded.

Part 2: Explore EXEC Modes

In this part of the activity, you will switch to privileged EXEC mode and issue additional commands.

Step 1:   Enter privileged EXEC mode.

a. At the prompt, type the question mark (?).

`S1> ?`

Question:

What information is displayed for the **enable** command?

_____

b. Type **en** and press the **Tab** key.

`S1> en<Tab>`

Question:

What displays after pressing the **Tab** key?

This is called command completion (or tab completion). When part of a command is typed, the **Tab** key can be used to complete the partial command. If the characters typed are enough to make the command unique, as in the case of the **enable** command, the remaining portion of the command is displayed.

Question:

What would happen if you typed **te<Tab>** at the prompt?

---

c. Enter the **enable** command and press **ENTER**.

Question:

How does the prompt change?

---

d. When prompted, type the question mark (**?**).

`S1# ?`

One command starts with the letter 'C' in user EXEC mode.

Question:

How many commands are displayed now that privileged EXEC mode is active? (**Hint:** you could type c? to list just the commands beginning with 'C'.)

Step 2: Enter Global Configuration mode.

a. When in privileged EXEC mode, one of the commands starting with the letter 'C' is **configure**. Type either the full command or enough of the command to make it unique. Press the **<Tab>** key to issue the command and press **ENTER**.

`S1# configure`

Question:

What is the message that is displayed?

---

b. Press **ENTER** to accept the default parameter that is enclosed in brackets [terminal].

Question:

How does the prompt change?

---

c. This is called global configuration mode. This mode will be explored further in upcoming activities and labs. For now, return to privileged EXEC mode by typing **end**, **exit**, or **Ctrl-Z**.

```
S1(config)# exit
S1#
```

Part 3: Set the Clock

Step 1:   Use the clock command.

a. Use the **clock** command to further explore Help and command syntax. Type **show clock** at the privileged EXEC prompt.

```
S1# show clock
```

Question:

What information is displayed? What is the year that is displayed?

_____

b. Use the context-sensitive help and the **clock** command to set the time on the switch to the current time. Enter the command **clock** and press **ENTER**.

```
S1# clock<ENTER>
```

Question

What information is displayed?

_____

c. The "% Incomplete command" message is returned by the IOS. This indicates that the **clock** command needs more parameters. Any time more information is needed, help can be provided by typing a space after the command and the question mark (?).

```
S1# clock ?
```

Question:

What information is displayed?

_____

d. Set the clock using the **clock set** command. Proceed through the command one step at a time.

```
S1# clock set ?
```

Questions:

What information is being requested?

_____

What would have been displayed if only the **clock set** command had been entered, and no request for help was made by using the question mark?

_____

e. Based on the information requested by issuing the **clock set ?** command, enter a time of 3:00 p.m. by using the 24-hour format of 15:00:00. Check to see if more parameters are needed.

```
S1# clock set 15:00:00 ?
```

The output returns a request for more information:

```
<1-31>  Day of the month
MONTH   Month of the year
```

f. Attempt to set the date to 01/31/2035 using the format requested. It may be necessary to request additional help using context-sensitive help to complete the process. When finished, issue the **show clock** command to display the clock setting. The resulting command output should display as:

```
S1# show clock
*15:0:4.869 UTC Wed Jan 31 2035
```

g. If you were not successful, try the following command to obtain the output above:

```
S1# clock set 15:00:00 31 Jan 2035
```

Step 2: Explore additional command messages.

a. The IOS provides various outputs for incorrect or incomplete commands. Continue to use the **clock** command to explore additional messages that may be encountered as you learn to use the IOS.

b. Issue the following commands and record the messages:

```
S1# cl<enter>
```

Questions:

What information was returned?

_____

Question:

What information was returned?

_____

Question:

What information was returned?

_____

Question:

What information was returned?

_____

c. Close **Configuration** window.

18.3.3 Packet Tracer–Use Cisco IOS Show Commands

Objectives

- Use Cisco IOS **show** commands.

Background / Scenario

The Cisco IOS **show** commands are used extensively when working with Cisco equipment. In this activity, you will use the **show** commands on a router located at an ISP.

Instructions

Part 1: Connect to the ISP Cisco 4321 Router

In this part, you will use the terminal emulation software on ISP PC to connect to the Cisco 4321 router.

   a. Click **ISP PC.**

   b. Click **Desktop** tab. Select **Terminal.** Review the terminal configuration and click **OK** to continue.

   c. The **ISPRouter >** prompt indicates that you are in user EXEC mode. Press **ENTER** if the prompt did not display.

Part 2: Explore the Show Commands

Use the information displayed by these **show** commands to answer the following questions.

Step 1:   Explore **show** commands in User EXEC mode.

   Open configuration window

   a. Type **show ?** at the prompt.

   Question:

   List a few more **show** commands that are available in the user EXEC mode.

   _____

   b. Enter **show arp** at the prompt.

   Question:

   Record the MAC address and the IP address listed.

   _____

   c. Enter **show flash** at the prompt.

   Question:

   Record the IOS image listed:

   _____

d. Enter **show ip route** at the prompt.

Question:

How many routes are listed in the table?

e. Enter **show interfaces** at the prompt.

Question:

Which interface is up and running?

Interface	Status	Protocol
GigabitEthernet 0/0/0	Up	
GigabitEthernet 0/0/1		Down
Serial0/1/0		
Serial0/1/1	Down	

f. Enter **show ip interface** at the prompt.

Question:

According to the **show ip interface** output, which interface is connected?

g. Enter **show version** at the prompt.

Question:

What technology package is enabled currently on the router?

h. Enter **show protocols** at the prompt.

Question:

Which protocols are enabled currently on the router?

i. Enter **show running-config** at the prompt.

Question:

What is the output?

Step 2:   Explore the **show** commands in privileged EXEC mode.

a. Enter **enable** at the prompt to enter the privileged EXEC mode.

Question:

List some additional **show** commands in this mode.

b. Enter **show running-config** at the prompt.

Question:

What is the output?

_____

c. Close **Configuration** window.

CHAPTER 19

Build a Small Cisco Network

Packet Tracer
☐ Activity

19.1.4 Packet Tracer–Implement Basic Connectivity

Addressing Table

Device	Interface	IP Address	Subnet Mask
S1	VLAN 1	192.168.1.253	255.255.255.0
S2	VLAN 1	192.168.1.254	255.255.255.0
PC1	NIC	192.168.1.1	255.255.255.0
PC2	NIC	192.168.1.2	255.255.255.0

Objectives

- Part 1: Perform a Basic Configuration on S1 and S2.
- Part 2: Configure the PCs.
- Part 3: Configure the Switch Management Interface.

Background / Scenario

In this activity, you will first perform basic switch configurations. Then you will implement basic connectivity by configuring IP addressing on the switches and PCs. When the IP addressing configuration is complete, you will use various **show** commands to verify configurations and use the ping command to verify basic connectivity between devices.

Instructions

Part 1: Perform SVI Configuration on S1 and S2

Step 1: Configure S1 with a hostname.

    a. Click **S1**, and then click the **CLI** tab.

    b. Enter the privileged EXEC mode. Then enter the global configuration mode.

```
Switch> enable
Switch# configure terminal
Enter configuration commands, one per line.  End with CNTL/Z.
```

    c. Configure the hostname as **S1**.

```
Switch(config)# hostname S1
S1(config)#
```

Step 2: Configure S1 with an IP address.

    Switches can be used without any configurations. Switches forward information from one port to another based on Media Access Control (MAC) addresses.

Question:

Why does a switch need an IP address?

_____

a. In the global configuration mode, enter the following commands to configure S1 with an IP address in VLAN 1.

```
S1(config)# interface vlan 1
S1(config-if)# ip address 192.168.1.253 255.255.255.0
S1(config-if)# no shutdown
%LINEPROTO-5-UPDOWN: Line protocol on Interface Vlan1, changed state
to up
```

Question:

What does the **no shutdown** command do?

_____

b. Exit the configuration mode and save the configuration.

```
S1(config-if)# end
S1#
S1# copy running-config startup-config
Destination filename [startup-config]?
Building configuration...
[OK]
```

c. Verify the IP address configuration on S1.

```
S1# show ip interface brief
<output omitted>
 Vlan1                   192.168.1.253   YES manual up                    up
```

Step 3:   Configure S2 with a hostname and IP address.

a. Click **S2**. In the CLI tab, enter the global configuration mode.

b. Configure the switch **S2** with a hostname using the information according to the Addressing Table.

c. Using the information in the Addressing Table, repeat the same process to configure the switch **S2** with an IP address.

d. Exit the configuration mode. Verify the IP address configuration on S2.

```
S2# show ip interface brief
<output omitted>
 Vlan1                   192.168.1.254   YES manual up                    up
```

e. Save the configuration file to NVRAM. Enter the **copy running-config startup-config** command to save the configuration.

Part 2: Configure the PCs

In this part, you will configure PC1 and PC2 with IP addresses and verify network connectivity.

Step 1:   Configure both PCs with IP addresses.

a. Click **PC1**, and then click the **Desktop** tab.

b. Click **IP Configuration.** In the **Addressing Table** above, you can see that the IP address for PC1 is supposed to be **192.168.1.1** and the subnet mask **255.255.255.0.** Enter this information for PC1 in the **IP Configuration** window.

c. Repeat the previous steps for PC2. Use the IP address listed in the Address Table for PC2.

Step 2:   Test connectivity from the PCs.

a. Click **PC1.** Close the **IP Configuration** window if it is still open. In the **Desktop** tab, click **Command Prompt.**

b. Enter the **ping** command and the IP address for S1.

```
Packet Tracer PC Command Line 1.0
PC> ping 192.168.1.253
```

Question:

Were you successful? Explain.

_____

c. From PC1, ping S2 and PC2.

d. Repeat the pings to S1, S2, and PC1 from PC2.

All pings should be successful. If your first ping result is 80%, retry; it should now be 100%. You will learn why a ping may fail the first time later in your studies. If you are unable to ping any of the devices, check your configuration for errors.

Step 3:   Verify network connectivity from the switches.

Network connectivity can be verified using the **ping** command. It is very important that connectivity exists throughout the network.

a. From **S1,** ping the other devices in the network. The ping to PC1 is displayed below as an example.

```
S1> ping 192.168.1.1

Type escape sequence to abort.
Sending 5, 100-byte ICMP Echos to 192.168.1.1, timeout is 2 seconds:
!!!!!
Success rate is 100 percent (5/5), round-trip min/avg/max = 0/0/1 ms
```

b. From S2, ping the other devices in the network.

All pings should be successful. If your first ping result is 80%, retry; it should now be 100%. You will learn why a ping may fail the first time later in your studies. If you are unable to ping any of the devices, check your configuration for errors.

Packet Tracer
☐ Activity

19.2.4 Packet Tracer–Configure Initial Router Settings

Objectives

- Part 1: Verify the Default Router Configuration.
- Part 2: Configure and Verify the Initial Router Configuration.
- Part 3: Save the Running Configuration File.

Background / Scenario

In this activity, you will perform basic router configuration tasks. You will secure access to the CLI and console port using encrypted and plaintext passwords. You will also configure messages for users who are logging into the router. These banners warn unauthorized users that access is prohibited. Finally, you will verify and save your running configuration.

Instructions

Part 1: Verify the Default Router Configuration

Step 1:   Establish a console connection to R1.

    a.   Choose a **Console** cable from the available connections.

    b.   Click **PCA** and select **RS 232**.

    c.   Click **R1** and select **Console**.

    d.   Click **PCA > Desktop** tab > **Terminal**.

    e.   Click **OK** and press **ENTER**. You are now able to configure **R1**.

Step 2:   Enter privileged mode and examine the current configuration.

You can access all the router commands from privileged EXEC mode. However, because many of the privileged commands configure operating parameters, privileged access should be password-protected to prevent unauthorized use.

    a.   Enter privileged EXEC mode by entering the **enable** command.

```
Router> enable
Router#
```

Notice that the prompt changed in the configuration to reflect privileged EXEC mode.

    b.   Enter the **show running-config** command.

```
Router# show running-config
```

Questions:

What is the router's hostname?

_____

How many FastEthernet interfaces does the Router have?

_____

How many Gigabit Ethernet interfaces does the Router have?

How many Serial interfaces does the router have?

What is the range of values shown for the vty lines?

c. Display the current contents of NVRAM.

```
Router# show startup-config
startup-config is not present
```

Question:

Why does the router respond with the **startup-config is not present** message?

Part 2: Configure and Verify the Initial Router Configuration

To configure parameters on a router, you may be required to move between various configuration modes. Notice how the prompt changes as you navigate through the IOS configuration modes.

Step 1: Configure the initial settings on R1.

Note: If you have difficulty remembering the commands, refer to the content for this topic.

a. Configure **R1** as the hostname.

b. Configure Message of the day text: **Unauthorized access is strictly prohibited.**

c. Encrypt all plaintext passwords.

Use the following passwords:

1) Privileged EXEC, encrypted: **itsasecret**

2) Console: **letmein**

Step 2: Verify the initial settings on **R1**.

a. Verify the initial settings by viewing the configuration for **R1**.

Question:

What command do you use?

b. Exit the current console session until you see the following message:

```
R1 con0 is now available

Press RETURN to get started.
```

c.   Press **ENTER**; you should see the following message:

```
Unauthorized access is strictly prohibited.

User Access Verification

Password:
```

Questions:

Why should every router have a message-of-the-day (MOTD) banner?

_____

If you are not prompted for a password before reaching the user EXEC prompt, what console line command did you forget to configure?

_____

d.   Enter the passwords necessary to return to privileged EXEC mode.

Questions:

If you configure any more passwords on the router, are they displayed in the configuration file as plaintext or in encrypted form? Explain.

_____

Part 3: Save the Running Configuration File

Step 1:   Save the configuration file to NVRAM.

a.   You have configured the initial settings for **R1**. Now back up the running configuration file to NVRAM to ensure that the changes made are not lost if the system is rebooted or loses power.

Questions:

What command did you enter to save the configuration to NVRAM?

_____

What is the shortest, unambiguous version of this command?

_____

Which command displays the contents of the NVRAM?

_____

b.   Verify that all the parameters configured are recorded. If not, analyze the output and determine which commands were not executed or were entered incorrectly. You can also click **Check Results** in the instruction window.

Step 2:   Optional: Save the startup configuration file to flash.

Although you will be learning more about managing the flash storage in a router in later chapters, you may be interested to know that, as an added backup procedure, you can save your startup configuration file to flash. By default, the router still loads the startup configuration from NVRAM, but if NVRAM becomes corrupt, you can restore the startup configuration by copying it over from flash.

Complete the following steps to save the startup configuration to flash.

a. Examine the contents of flash using the **show flash** command:

```
R1# show flash
```

Questions:

How many files are currently stored in flash?

_____

Which of these files would you guess is the IOS image?

_____

Why do you think this file is the IOS image?

_____

b. Save the startup configuration file to flash using the following commands:

```
R1# copy startup-config flash
Destination filename [startup-config]
```

The router prompts you to store the file in flash using the name in brackets. If the answer is yes, then press **ENTER**; if not, type an appropriate name and press **ENTER**.

c. Use the **show flash** command to verify the startup configuration file is now stored in flash.

19.3.6 Packet Tracer–Configure SSH

Addressing Table

Device	Interface	IP Address	Subnet Mask
S1	VLAN 1	10.10.10.2	255.255.255.0
PC1	NIC	10.10.10.10	255.255.255.0

Objectives

- Part 1: Secure Passwords.

- Part 2: Encrypt Communications.

- Part 3: Verify SSH Implementation.

Background / Scenario

SSH should replace Telnet for management connections. Telnet uses insecure plaintext communications. SSH provides security for remote connections by providing strong encryption of all transmitted data between devices. In this activity, you will secure a remote switch with password encryption and SSH.

Instructions

Part 1: Secure Passwords

a. Using the command prompt on **PC1**, Telnet to **S1**. The user EXEC and privileged EXEC password is **cisco**.

```
PC> telnet 10.10.10.2
Trying 10.10.10.2 ...Open


User Access Verification

Password:
S1> en
Password:
S1#
```

b. Save the current configuration so that any mistakes you might make can be reversed by toggling the power for **S1**.

```
S1# copy running-config startup-config
Destination filename [startup-config]?
Building configuration...
[OK]
```

c. Show the current configuration and note that the passwords are in plaintext.

d. In the global configuration mode, enter the command that encrypts plaintext passwords:

```
S1(config)# service password-encryption
```

e. Verify that the passwords are encrypted.

Part 2: Encrypt Communications

Step 1: Set the IP domain name and generate secure keys.

It is generally not safe to use Telnet, because data is transferred in plaintext. Therefore, use SSH whenever it is available.

a. Configure the domain name to be **netacad.pka**.

```
S1(config)# ip domain-name netacad.pka
```

b. Secure keys are needed to encrypt the data. Generate the RSA keys using a 1024 key length.

```
S1(config)# crypto key generate rsa
The name for the keys will be: S1.netacad.pka
Choose the size of the key modulus in the range of 360 to 2048 for your
    General Purpose Keys. Choosing a key modulus greater than 512 may take
    a few minutes.

How many bits in the modulus [512]: 1024
% Generating 1024 bit RSA keys, keys will be non-exportable...[OK]
```

Step 2: Create an SSH user and reconfigure the vty lines for SSH-only access.

a. Create an **administrator** user with **cisco** as the secret password.

```
S1(config)# username administrator secret cisco
```

b. Configure the vty lines to check the local username database for login credentials and to only allow SSH for remote access. Remove the existing vty line password.

```
S1(config)# line vty 0 15
S1(config-line)# login local
S1(config-line)# transport input ssh
S1(config-line)# no password cisco
```

Part 3: Verify SSH Implementation

a. Exit the Telnet session and attempt to log back in using Telnet. The attempt should fail.

b. Attempt to log in using SSH. Type **ssh** and press **ENTER** without any parameters to reveal the command usage instructions. Hint: The **-l** option is the letter "L", not the number 1.

c. Upon successful login, enter privileged EXEC mode and save the configuration. If you were unable to successfully access **S1**, toggle the power and begin again at Part 1.

19.4.4 Packet Tracer–Build a Switch and Router Network

Addressing Table

Device	Interface	IP Address	Subnet Mask	Default Gateway
R1	G0/0/0	192.168.0.1	255.255.255.0	N/A
	G0/0/1	192.168.1.1	255.255.255.0	N/A
S1	VLAN 1	192.168.1.2	255.255.255.0	192.168.1.1
PC-A	NIC	192.168.1.3	255.255.255.0	192.168.1.1
PC-B	NIC	192.168.0.3	255.255.255.0	192.168.0.1

Objectives

Part 1: Configure Devices and Verify Connectivity.

- Assign static IP information to the PC interfaces.

- Configure the router and switch.

- Verify network connectivity.

Part 2: Display Device Information.

- Retrieve hardware and software information from the network devices.

- Interpret the output from the routing table.

- Display interface information on the router.

- Display a summary list of the interfaces on the router and switch.

Part 3: Secure Remote Access to the Router.

- Set the IP domain name and generate secure keys.

- Create an SSH user and configure vty lines for SSH-only access.

- Verify SSH implementation.

Background / Scenario

In this lab, you will cable the equipment and configure the devices to match the Addressing Table. After the configurations have been saved, you will verify your configurations by testing for network connectivity.

After the devices have been configured and network connectivity has been verified, you will use IOS commands to retrieve information from the devices to answer questions about your network equipment. You will also access the router remotely via SSH.

Instructions

Part 1: Configure Devices and Verify Connectivity

In Part 1, you will set up the network topology and configure basic settings, such as the interface IP addresses, device access, and passwords. Refer to the Addressing Table at the beginning of this activity for device names and address information.

Step 1:   Connect the devices.

The devices are already deployed in the workspace. You will connect them using the correct cables between the devices as listed below:

- Connect **PCA F0** to **S1 F0/1**.
- Connect **S1 G0/1** to **R1 G0/0/1**.
- Connect **R1 G0/0/0** to **PCB F0**.

Step 2:   Assign static IP information to the PC interfaces.

    a.   Configure the IP address, subnet mask, and default gateway settings on PC-A.

    b.   Configure the IP address, subnet mask, and default gateway settings on PC-B.

    c.   Ping PC-B from a command prompt window on PC-A.

    Question:

    Why were the pings not successful?

Step 3:   Configure **R1**.

    a.   Console into the router and enable privileged EXEC mode. (**Hint:** Use console cable and terminal on a PC.)

    b.   Enter configuration mode.

    c.   Assign a device name to the router according to the Addressing Table.

    d.   Assign **class** as the privileged EXEC encrypted password.

    e.   Assign **cisco** as the console password and enable login.

    f.   Encrypt the plaintext passwords.

    g.   Create a banner that warns anyone accessing the device that unauthorized access is prohibited.

    h.   Configure the IP addresses according to the Addressing Table and activate both Ethernet interfaces on the router.

    i.   Save the running configuration to the startup configuration file.

    Question:

    Were the pings successful? Explain.

Step 4:   Configure **S1**.

Note: Most of the commands on the switch are similar to the commands on the router in this step. Use the **help (?)** context as necessary.

    a.   Console into the switch and enable privileged EXEC mode.

    b.   Enter configuration mode.

    c.   Assign a device name to the switch according to the Addressing Table.

    d. Assign **class** as the privileged EXEC encrypted password.

    e. Assign **cisco** as the console password and enable login.

    f. Encrypt the plaintext passwords.

    g. Create a banner that warns anyone accessing the device that unauthorized access is prohibited.

    h. Configure the IP address for the SVI for VLAN 1 according to the Addressing Table and activate the interface.

    i. Configure the default gateway according to the Addressing Table.

    j. Save the running configuration to the startup configuration file.

Part 2: Display Device Information

Step 1: Retrieve hardware and software information from the network devices.

    a. Use the **show version** command to answer the following questions about the router.

```
R1# show version
```
Question:

What is the name of the IOS image that the router is running?

_____

    b. Use the **show version** command to answer the following questions about the switch.

```
S1# show version
```
Questions:

What are the IOS software image and version running on the switch?

_____

What is the model number of the switch?

_____

Step 2: Display the routing table on the router.

Use the **show ip route** command on the router to answer the following questions.

```
R1# show ip route
```
Questions:

What code is used in the routing table to indicate a directly connected network?

_____

How many route entries are coded with a C code in the routing table?

_____

What interface types are associated to the C coded routes?

_____

Step 3:   Display interface information on the router.

Use the **show interface** g0/0/1 to answer the following questions.

`R1# show interfaces g0/0/1`

Questions:

What is the operational status of the G0/0/1 interface?

_____

What is the Media Access Control (MAC) address of the G0/01 interface?

_____

How is the Internet address displayed in this command?

_____

Step 4:   Display a summary list of the interfaces on the router and switch.

There are several commands that can be used to verify an interface configuration. One of the most useful of these is the **show ip interface brief** command. The command output displays a summary list of the interfaces on the device and provides immediate feedback to the status of each interface.

a.   Enter the **show ip interface brief** command on the router.

```
R1# show ip interface brief
Interface             IP-Address      OK? Method Status              Protocol
GigabitEthernet0/0/0  192.168.0.1     YES NVRAM  up                  up
GigabitEthernet0/0/1  192.168.1.1     YES NVRAM  up                  up
Serial0/1/0           unassigned      YES unset  down                down
Serial0/1/1           unassigned      YES unset  down                down
Vlan1                 unassigned      YES NVRAM  administratively     down
                                                 down
```

b.   Enter the **show ip interface brief** command on the switch.

```
S1# show ip interface brief
Interface             IP-Address      OK? Method Status              Protocol
FastEthernet0/1       unassigned      YES unset  down                down
FastEthernet0/2       unassigned      YES unset  down                down
<output omitted>
GigabitEthernet0/1    unassigned      YES unset  up                  up
GigabitEthernet0/2    unassigned      YES unset  down                down
Vlan1                 192.168.1.2     YES manual up                  up
```

Part 3: Secure Remote Access to the Router

Step 1:   Set the IP domain name and generate secure keys.

a.   On **R1**, configure the domain name as **academy.net**.

```
R1(config)# ip domain-name academy.net
```

    b. Generate RSA keys with a **1024** key length.

```
R1(config)# crypto key generate rsa
The name for the keys will be: R1.academy.net
Choose the size of the key modulus in the range of 360 to 2048 for your
   General Purpose Keys. Choosing a key modulus greater than 512 may
   take a few minutes.

How many bits in the modulus [512]: 1024
% Generating 1024 bit RSA keys, keys will be non-exportable...[OK]
```

Step 2: Create an SSH user and configure vty lines for SSH-only access.

    a. Create a user with **SSHuser** as the username and **cisco** as the secret password.

```
R1(config)# username SSHuser secret cisco
```

    b. Configure the vty lines to use the local username database for login credentials.

```
R1(config)# line vty 0 4
R1(config-line)# login local
```

    c. The vty lines should only allow SSH for remote access.

```
R1(config-line)# transport input ssh
```

Step 3: Verify SSH Implementation.

    a. Click PCA, select **Command Prompt** in the Desktop tab.

    b. At the prompt, enter **ssh -l SSHuser 192.168.1.1.**

    c. Enter **cisco** when prompted for the password.

    Question:

    What is the displayed message?

    _____

    You should be at the prompt of R1. If you are not successful, verify the configurations are correct and the credentials were entered correctly.

Reflection

  1. If the G0/0/1 interface showed administratively down, what interface configuration command would you use to turn the interface up?

    _____

  2. What would happen if you had incorrectly configured interface G0/0/1 on the router with an IP address of 192.168.1.2?

    _____

CHAPTER 20

Security Considerations

20.3.3 Packet Tracer–Use the ipconfig Command

Objectives

- Use the ipconfig command to identify incorrect configuration on a PC.

Background / Scenario

A small business owner cannot connect to the internet with one of the four PCs in the office. All the PCs are configured with static IP addressing using the **192.168.1.0 /24** network. The PCs should be able to access **www.cisco.pka** webserver. Use the **ipconfig /all** command to identify which PC is incorrectly configured.

Instructions

Part 1: Verify Configurations

a. Access the **Command Prompt** on each PC and type the command **ipconfig /all** at the prompt.

b. Examine the IP address, subnet mask, and default gateway configuration for each PC. Be sure to record this IP configuration for each PC to help identify any PCs that are incorrectly configured.

Part 2: Correct Any Misconfigurations

a. Select the PC that is incorrectly configured.

b. Click the **Desktop** tab > **IP Configuration** tab to correct the misconfiguration.

20.3.6 Packet Tracer–Use the ping Command

Objectives

- Use the **ping** command to identify an incorrect configuration on a PC.

Background / Scenario

A small business owner learns that some users are unable to access a website. All PCs are configured with static IP addressing. Use the **ping** command to identify the issue.

Instructions

Part 1: Verify Connectivity

Access the **Desktop** tab > **Web Browser** of each PC and enter the URL **www.cisco.pka**. Identify any PCs that are not connecting to the web server.

Note: All the devices require time to complete the boot process. Please allow up to one minute before receiving a web response.

Question:

Which PCs are unable to connect to the web server?

Part 2: Ping the Web Server from PCs with Connectivity Issues

a. On the PC, access the **Command Prompt** from the **Desktop** tab.

b. At the prompt, enter **ping www.cisco.pka**.

Question:

Did the ping return a reply? What is the IP address displayed in the reply, if any?

Part 3: Ping the Web Server from Correctly Configured PCs

a. On the PC, access the **Command Prompt** from the **Desktop** tab.

b. At the prompt, enter **ping www.cisco.pka**.

Question:

Did the **ping** return a reply? What is the IP address returned, if any?

Part 4: Ping the IP Address of the Web Server from PCs with Connectivity Issues

a. On the PC, access the **Command Prompt** from the **Desktop** tab.

b. Attempt to reach the IP address of the web server with the **ping** command.

Did the **ping** return a reply? If so, then the PC can reach the web server via IP address, but not domain name. This could indicate a problem with the DNS server configuration on the PC.

Part 5: Compare the DNS Server Information on the PCs

a. Access the **Command Prompt** of the PCs without any issues.

b. Using the command **ipconfig /all**, examine the DNS server configuration on the PCs without any issues.

c. Access the **Command Prompt** of the PCs with connectivity issues.

d. Using the command **ipconfig /all**, examine the DNS server configuration on the PCs with misconfigurations. Do the two configurations match?

Part 6: Make Any Necessary Configuration Changes on the PCs

a. Navigate to the **Desktop** tab of the PCs with issues, and make any necessary configuration changes in **IP Configuration**.

b. Using the **Web Browser** within the **Desktop** tab, connect to **www.cisco.pka** to verify that the configuration changes resolved the problem.

20.3.12j Lab–Troubleshoot Using Network Utilities

Objectives

- Interpret the output of commonly used network command line utilities.

- Determine which network utility can provide the necessary information to perform troubleshooting activities in a bottom-up troubleshooting strategy.

Background / Scenario

There are a number of problems that can cause networking connectivity issues. In this lab, you will use network utilities that can help you to identify connectivity issues in wireless networks. The network command line utilities are also useful to detect problems in a wired network.

Required Resources

- 1 PC (Windows 10 with a wired and wireless NIC installed)

- A wireless router

- Internet connectivity

Instructions

Part 1: Network Connections

Step 1:   Connect to a wireless network.

   a. Disconnect the Ethernet cable from your computer. An "orange triangle" appears over the **Connections** icon in the system tray.

   b. Click the **Connection** icon in the system tray.

   Question:

   What is the name of an available wireless connection?

   _____

   a. Click one of the available wireless connections. Connect to the network. Enter login information if required.

   b. Confirm that the connection is successful.

Step 2:   Verify that the network adapter is operational.

   When a connectivity problem is reported, the first step in a bottom-up trouble-shooting strategy is to determine whether the NIC and the operating system settings on the computer are functioning correctly.

   a. Open the Control Panel and select **Network and Sharing Center**. Right-click **Start** and select **Control Panel**. Click **Network and Sharing Center**. Click **Change adapter settings**.

b. Select the **Wireless Network** connection. Right click the adapter and select **Status** from the menu. If the **Status** choice is grayed out, it indicates that the adapter is either not enabled or not connected to a wireless SSID.

c. In the status window, verify that the connection is enabled and that the connection SSID is correct. Click **Details** to open the adapter details window.

d. The **Details** window shows the current IP configuration active on the network adapter. It displays both the IPv4 and IPv6 configurations. If DHCP is active, the lease information is shown.

Questions:

Is DHCP enabled on the PC?

When does the DHCP lease expire?

Step 3: Confirm correct network configuration.

a. Open a **Command Prompt**.

b. Enter **ping 127.0.0.1**. The IP address 127.0.0.1 is also referred to as the localhost address. A successful ping to the localhost address indicates that the TCP/IP protocol stack is operational on the computer. If the localhost address does not reply to a ping command, there might be an issue with the device driver or the network interface card.

Question:

Was the **ping** command successful?

c. Use the **ipconfig** command. Identify the IP address, subnet mask and default gateway addresses configured on the computer.

If the local IPv4 address is a host address on the 169.254.0.0/16 network, the computer received its IP address configuration through the Automatic Private IP Addressing (APIPA) feature of the Windows operating system.

Question:

What problems can cause a computer to receive an APIPA address?

If the computer is assigned an APIPA address, there might be an issue with the DHCP server. If the wireless router is providing the DHCP services, confirm that the DHCP service is configured correctly and that the IP address range is large enough to accommodate all the devices that may attach wirelessly.

Question:

What is the IP address of the default gateway assigned to your PC?

d. To test if the PC can reach the default gateway through the network, ping the default gateway IP address.

A successful ping indicates that there is a connection between the computer and the default gateway.

If the **ping** command does not complete successfully, make sure that the IP address of the gateway is typed correctly and that the wireless connection is active.

e. Type **net view**. The **net view** command, when issued on a Windows PC, displays the computer names of other Windows devices in your Windows domain or workgroup. When **net view** displays the names of other computers it indicates that your computer can successfully send messages across the network.

```
C:\Users\Student> net view
```

Question:

List the computer names that are displayed.

_____

Note: Depending on the configuration of the PCs in your lab, **net view** may not return any computer names or may display an error message. If this is the case, move on to the next step.

Part 2: External Connectivity

Step 1:   Test external connectivity.

If you have an external connection, use the following methods to verify the operation of the default gateway and the DNS service.

Note: Your output may be varied.

a. The Windows **tracert** command performs the same function as the **traceroute** command used within the Cisco IOS. Use the **tracert** command along with your school's website URL or the Cisco Networking Academy website. For example, enter **tracert www.netacad.com**.

```
C:\Users\Student> tracert www.netacad.com

Tracing route to Liferay-Prod-1009279580.us.-east-1.elb.amazonaws.com
[52.5.233.103]

over a maximum of 30 hops:
    1     1 ms     57 ms      3 ms   192.168.1.1
    2     *        12 ms     12 ms   10.39.176.1
    3    14 ms     28 ms     11 ms   100.127.65.248
    4    10 ms     26 ms     21 ms   70.169.73.90
    5    35 ms     32 ms     36 ms   68.1.2.109
<output omitted>
Trace complete.
```

The **tracert** command displays the path that the packet takes between the source and destination IP addresses. Each router that the packet travels through to reach the destination address is shown as a hop in the **tracert** output. If there is a network issue on the path, the **tracert** output will stop after the last successful hop. The first hop in the output is the default gateway of the source PC, and the last entry will be the destination address when the **tracert** command completes successfully.

b. The command **tracert** uses the configured DNS server to resolve the fully qualified domain name to an IP address before beginning to trace the router to the destination. Using **tracert** or **ping** with a domain name instead of an IP address can confirm that the DNS server is providing name resolution services.

Questions:

What IP address was returned by the DNS server?

_____

What would happen if the DNS server could not resolve the domain name of the server?

_____

c. Use the **nslookup** command with the IP address you just discovered. The **nslookup** command is a utility that can be used to troubleshoot DNS problems.

Type **nslookup 72.163.6.233**. The IP address in this example is assigned to a server at Cisco Systems.

Questions:

What domain name was returned?

_____

What DNS server did the **nslookup** command use to resolve the domain name?

_____

Does the DNS server IP address match the one displayed in the **ipconfig /all** output?

_____

d. When the configured DNS server cannot resolve domain names or IP addresses, it is possible to set **nslookup** to try to resolve the names using a different DNS server. If another DNS server can resolve the addresses, but the configured DNS server cannot, there could be a problem with the DNS server configuration. Enter **nslookup /?** to view the options that can be used to test and troubleshoot DNS issues.

```
C:\Users\Student> nslookup /?
Usage:
    nslookup [-opt ...]            # interactive mode using default
                                     server
    nslookup [-opt ...] - server  # interactive mode using 'server'
    nslookup [-opt ...] host      # just look up 'host' using default
                                     server
    nslookup [-opt ...] host server # just look up 'host' using 'server'
```

Step 2:   Test Application layer connectivity.

Open a web browser. Enter **www.cisco.com** in the Address field.

Question:

Does the Cisco.com web page load in the browser? What underlying network functions have to be working for the web page load?

_____

Reflection

1. The steps in this lab represent a bottom-up troubleshooting strategy, where the effort starts with the OSI model physical layer and finishes with verifying the functionality of the application layer. What are the other two troubleshooting strategies used by network technicians to isolate problems?

_____

2. Which strategy would you try first when presented with a network connectivity problem? Explain.

_____

20.4.3 Packet Tracer–Troubleshoot a Wireless Connection

Objectives

- Identify and correct any misconfiguration of a wireless device.

Background / Scenario

A small business owner learns that a wireless user is unable to access the network. All the PCs are configured with static IP addressing. Identify and resolve the issue.

Instructions

Part 1: Verify Connectivity

Access the **Desktop > Web Browser** of each wireless PC and type **www.cisco.pka** into the URL. Identify any PCs that are not connecting to the web server.

Note: All the devices require time to complete the boot process. Please allow up to one minute before receiving a web response.

Question:

Which wireless PCs are unable to connect to the web server?

Part 2: Examine the IP Configuration of the PCs

a. On the PC that is unable to connect, access the **Command Prompt** from the **Desktop** tab.

b. Enter the **ipconfig /all** command at the prompt.

Question:

What IP addressing information is available?

Part 3: Examine the Wireless Settings on the Wireless Client

a. On the **Desktop** tab of any PC that is unable to connect, click **PC Wireless** to access the wireless configurations.

b. Click the **Connect** tab and record the associated SSID. Click **Refresh** as needed to display the list of SSIDs.

Question:

What is the associated SSID?

Part 4: Examine the Wireless Settings on the Wireless Router

a. Access the **Wireless Router** from the web browser of a wired PC. Use the username **admin** and password **admin** to access the wireless router.

Question:

What IP address did you use? (**Hint:** default gateway)

_____

b. On the **Basic Setup** page, examine the **DHCP Server Setting** configuration.

Question:

Is DHCP enabled?

_____

c. Click the **Wireless** tab.

d. Examine the setup information under the **Wireless** tab.

Question:

What is the SSID? Does it match the SSID configured on the client?

_____

e. Click the **Wireless Security** submenu.

f. Examine the security settings.

Question:

What is the wireless security mode? What is the passphrase?

_____

Part 5: Make Any Necessary Configuration Changes on the Wireless Clients

a. On the **Desktop** tab of any PC that is unable to connect, click **PC Wireless** to correct the wireless configurations.

b. Click the **Connect** tab. Select the **Academy** wireless network and click **Connect**.

c. Enter the passphrase (Pre-shared Key) recorded from the wireless router. Click **Connect**.

d. Using the web browser within the **Desktop** tab, connect to **www.cisco.pka** to verify that the configuration changes resolved the problem.

Packet Tracer
☐ Activity

20.7.1 Packet Tracer–Skills Integration Challenge

Addressing Table

Device	Interface	IP Address	Subnet Mask	Default Gateway
R1	G0/0/1	209.165.201.1	255.255.255.224	N/A
S1	VLAN 1	209.165.201.2	255.255.255.224	N/A
Server	NIC	209.165.201.30	255.255.255.224	209.165.201.1

Objectives

- Configure IP addresses.
- Set up wireless configuration in home network.
- Verify connectivity.

Background / Scenario

This activity includes many of the skills that you have acquired during your Networking Essentials studies. First, you will configure the IP addresses on network devices in a simplified network. Second, you will set up the wireless configurations in home network. Finally, you will verify your implementation by testing end-to-end connectivity by accessing the web server, www.server.pka, and router R1 using SSH in the simplified network.

Instructions

Router R1

- Configure the device name according to the Addressing Table.
- Configure the IP address on G0/0/1 interface according to the Addressing Table and enable the interface.
- Create a banner that warns anyone accessing the device that unauthorized access is prohibited. Make sure to include the word **warning** in the banner.
- Assign **cisco** as the console password and enable login.
- Assign **class** as the encrypted privileged EXEC mode password.
- Encrypt all plaintext passwords.

Configure SSH on R1:

- Set the domain name to **networking.pka**.
- Generate a **1024**-bit RSA key.
- Create a user with a username **admin** with a secret password **cisco123**.
- Configure the vty lines for SSH access.
- Use the local user profiles for authentication.

Switch S1

- Configure the device name according to the Addressing Table.
- Configure the IP address of the switch on SVI interface according to the Addressing Table and enable the interface.

Server

Configure the IP address of the server according to the Addressing Table.

Wireless Router in the Home

Enter the **Home** cluster. From the web browser on PC, configure the following:

Initial Wireless Router IP Address:	**192.168.1.1**
Username / Password:	**admin / admin**
SSID for 2.4 GHz:	**MyHome**
Security Mode for 2.4 GHz:	**WPA2 Personal**
Passphrase for 2.4 GHz:	**123Cisco**
For both 5GHz:	**Disabled**
DHCP Configuration:	
Wireless Router IP Address:	**192.168.20.1**
Starting IP Address:	**192.168.20.101**
Maximum Number:	**100**
DNS 1:	**209.165.201.30**

End Devices in the Home

Configure the wireless settings so the end devices can access **www.server.pka**.

SSID:	**MyHome**
Security Mode:	**WPA2 Personal** or **WPA2-PSK**
Passphrase:	**123Cisco**

Note: For Tablet PC and Pda, use the Config tab for the wireless configurations.

Verify Connectivity

- Verify that IP addresses are in the correct networks. All the end devices should be in 192.168.20.0/24 network. If they are not in the correct network, enter the following commands at the command prompt.

  ```
  PC> ipconfig /release
  PC> ipconfig /renew
  ```

- Verify that all end devices in the Home can access **www.server.pka**.

- Verify that all end devices in the Home can access R1 via SSH with password **cisco123**.

  ```
  PC> ssh -l admin 209.165.201.1
  ```